The AI Investor

by

Alander L Pulliam

The AI Investor

Alander Lee Pulliam Jr and Olom "Nosa" Okonta

Published by Alander Lee Pulliam Jr, 2025.

THE AI INVESTOR

First edition. July 21, 2025.

ISBN: 979-8230388296

Written by Alander Lee Pulliam Jr and Olom "Nosa" Okonta.

Table of Contents

The AI Investor 1
Load the S&P 500 data 191
Calculate the 20-day SMA 192
Generate trading signals 193
Calculate daily returns 194
Calculate strategy returns (simplified, ignoring transaction costs and position sizing) 195
Calculate cumulative returns 196
Analyze results 197

To the dreamers who see beyond the horizon,and the innovators who dare to reshape the future—this book is for you.

To those who believe that knowledge is power,and that technology is a tool to elevate us all.

May you find the courage to embrace change,the curiosity to explore new possibilities,and the discipline to turn your visions into reality.

Remember, the journey to financial mastery begins with a single step—keep moving forward, stay inspired, and never stop learning.

This is your time to redefine what's possible.

With hope and belief in your limitless potential,Alander Pulliam

As I close the pages of this journey, I want you to remember that the future of investing is not just about algorithms or data—it's about you. It's about your mindset, your discipline, and your willingness to adapt. AI is a powerful tool, but it's your curiosity and commitment that will truly unlock your potential.

The world is changing faster than ever, and those who embrace innovation will find opportunities where others see obstacles. Keep learning, stay disciplined, and never underestimate your ability to shape your financial destiny.

This is only the beginning. The future belongs to those who dare to dream big and act boldly. I hope this book inspires you to take that first step—and to keep moving forward.

Here's to your journey ahead.

This book is dedicated to the intrepid pioneers of the financial technology revolution, the brilliant minds behind the algorithms and the dedicated individuals who tirelessly work to improve the accessibility and efficiency of investing. It is also dedicated to every aspiring investor who dares to dream of financial independence and seeks innovative tools and strategies to make those dreams a reality. May this book serve as a guide and an inspiration on your path toward achieving financial success and empowerment through the intelligent application of artificial intelligence. It's a testament to the power of human ingenuity and technological advancement in the pursuit of a more prosperous and secure financial future. To those who continually challenge the status quo, strive for better solutions, and are passionate about bridging the gap between technology and financial literacy: this work is for you.

The journey of investment strategies is a fascinating narrative of adaptation and innovation, a continuous evolution shaped by technological advancements, economic shifts, and the ever-present human desire for higher returns. Initially, investment decisions were largely driven by intuition, gut feeling, and perhaps a dash of insider information. This era, characterized by its manual and often opaque methods, saw investors relying heavily on fundamental analysis – painstakingly studying company financials, industry trends, and macroeconomic indicators to make informed (or, as often proved to be the case, uninformed) choices. Think of the early stockbrokers, poring over reams of paper, meticulously charting price movements with pencils and rulers. Their decisions were often based on limited data, subjective interpretations, and a significant element of chance. The results were naturally unpredictable, with market swings often driven by speculation and sentiment rather than robust analysis.

The limitations of relying solely on fundamental analysis became increasingly apparent, particularly during periods of rapid market volatility. The Great Depression, for instance, exposed the fragility of purely intuition-based investment strategies, leading to widespread financial ruin. The ensuing decades witnessed a gradual shift towards a more quantitative approach. The advent of computers in the mid-20th century enabled the processing of vast quantities of data, paving the way for the development of quantitative analysis. This marked a pivotal moment, as investors could now analyze historical market data, identify patterns, and develop sophisticated statistical models to predict future price movements. This quantitative revolution, fueled by the increasing power of computing, led to the emergence of quantitative investment strategies like statistical arbitrage and mean reversion.

However, even quantitative analysis faced limitations. The complexity of financial markets and the sheer volume of data involved quickly overwhelmed even the most powerful computers and sophisticated algorithms. Predicting market behavior with absolute certainty remained elusive, as unforeseen events and human emotions continued to influence market dynamics in unpredictable ways. The 1987 Black Monday crash, a stark reminder of market unpredictability, further highlighted the inadequacy of even the most advanced quantitative models in fully capturing market behavior. The crash showed that algorithms, however sophisticated, were not immune to sudden, dramatic shifts in market sentiment.

The subsequent decades witnessed the continuous refinement of quantitative strategies, alongside a growing emphasis on risk management. Modern portfolio theory (MPT), for instance, provided a framework for constructing diversified portfolios that maximized returns while minimizing risk. The development of advanced econometric techniques further improved the accuracy of market forecasts, leading to more sophisticated investment strategies. However, a critical challenge remained: how to efficiently process and interpret the increasingly massive and complex datasets that characterized the modern financial world? The answer, increasingly, lies in the realm of artificial intelligence.

The integration of AI into investment strategies marks the latest phase in this ongoing evolution. AI's ability to analyze colossal amounts of data, identify subtle patterns, and learn from past experiences has opened up unprecedented possibilities. Machine learning, a branch of AI, enables algorithms to learn from data without explicit programming. Deep learning, a subfield of machine learning, utilizes artificial neural networks with multiple layers to extract complex relationships and patterns from vast datasets, offering far greater predictive power than previous quantitative methods.

The application of AI has manifested in various forms. Robo-advisors, for example, utilize algorithms to automate portfolio construction, rebalancing, and even tax optimization. These platforms leverage AI to provide personalized investment advice, tailored to an individual's risk tolerance, financial goals, and investment horizon. Meanwhile, algorithmic trading employs AI-driven algorithms to execute trades at optimal prices and speeds, often outperforming human traders in terms of efficiency and speed of execution. Furthermore, AI-powered platforms are used for sentiment analysis, analyzing news articles, social media posts, and other textual data to gauge market sentiment and predict future price movements. The use of natural language processing (NLP) allows machines to understand and interpret human language, providing valuable insights into market sentiment and news events that might otherwise be missed by human analysts.

The evolution of investment strategies from manual methods to AI-driven approaches is not merely a technological advancement; it's a reflection of the ever-increasing complexity of financial markets. The shift demonstrates the ongoing effort to improve prediction accuracy, enhance efficiency, and mitigate risks. While AI promises significant advantages, it's crucial to acknowledge the potential pitfalls. Algorithmic bias, model inaccuracy, and data security concerns are significant challenges that need careful attention. The future success of AI in investing will depend on not only technological advancement but also on responsible development, robust risk management, and a clear understanding of ethical considerations. The journey continues, with AI playing an increasingly central role in shaping the future of investment strategies. But the core principles of disciplined investing—patience, diversification, risk management, and a long-term perspective—remain as essential as ever, serving as a robust foundation upon which to build a successful AI-powered investment strategy. The blend of human judgment and AI-powered tools will likely define the most successful approaches in the years to come. The human element, with its capacity for critical thinking, adaptability, and ethical decision-making, remains irreplaceable, even in an increasingly automated world.

Artificial intelligence (AI), a term often shrouded in science fiction narratives, is, at its core, about enabling machines to mimic human intelligence. This isn't about creating sentient robots, but rather about developing systems that can learn, reason, and solve problems – tasks previously considered the exclusive domain of humans. Think of AI as a powerful toolkit, offering a vast array of computational capabilities that can be applied across numerous fields, including, crucially, investment management.

At the heart of AI lies the concept of algorithms – sets of instructions that tell a computer how to perform a specific task. These algorithms aren't simply pre-programmed rules; many AI systems, particularly those leveraging machine learning, are capable of learning and adapting based on the data they are fed. This learning process is what differentiates AI from traditional computer programming. In traditional programming, you explicitly tell the computer exactly what to do, step by step. With AI, you provide the system with data, and it figures out the best way to achieve a desired outcome.

Machine learning (ML), a subset of AI, is where the real magic happens. Instead of relying on explicit programming, ML algorithms learn from data. Imagine teaching a child to identify a cat. You wouldn't give them a rigid set of rules (e.g., "a cat has four legs, whiskers, and a tail"). Instead, you'd show them numerous pictures of cats, pointing out their features. Over time, the child learns to identify cats based on patterns and characteristics, even if some cats look quite different from others. ML algorithms work in a similar way. They are fed large amounts of data, and they identify patterns and relationships within that data to make predictions or decisions.

Deep learning (DL), a more advanced form of ML, takes this process a step further. DL algorithms utilize artificial neural networks, inspired by the structure and function of the human brain. These networks consist of multiple layers of interconnected nodes, allowing the algorithm to learn highly complex patterns and relationships within data. The more layers a neural network has, the more intricate the patterns it can identify. Think of it like building upon that child's cat-identification skills. First, they learn to identify basic feline features. Then, with more experience, they can distinguish between different cat breeds, even if they've never seen that breed before. Deep learning allows AI to perform this level of nuanced pattern recognition.

To illustrate the practical applications of these concepts, let's consider a simple example: predicting stock prices. Traditional quantitative analysis might involve building a model based on a few key indicators, like price-to-earnings ratio and volume. However, this approach overlooks a vast amount of potentially relevant information – news articles, social media sentiment, economic reports, even weather patterns that could impact certain sectors. AI, and particularly deep learning, can analyze all this data simultaneously, uncovering subtle correlations and patterns that would be impossible for a human analyst to identify. It might reveal that a seemingly insignificant tweet about a company's product could foreshadow a future price drop, or that a specific weather pattern consistently correlates with increased demand for a particular commodity.

The ability of AI to handle massive, complex datasets is a game-changer in the world of investing. Human analysts, even with the most advanced tools, are limited in their capacity to process and interpret the sheer volume of data available today. AI can sift through terabytes of information in seconds, identifying subtle patterns and anomalies that could signal opportunities or risks. This allows for more informed and potentially more profitable investment decisions.

However, it's crucial to acknowledge that AI is not a magic bullet. AI systems are only as good as the data they are trained on. Biased or incomplete data will lead to biased or inaccurate predictions. Moreover, the "black box" nature of some AI algorithms – the difficulty in understanding exactly how they arrive at their conclusions – can be a concern. This lack of transparency can make it challenging to identify and address potential errors or biases in the system. This is a significant area of ongoing research and development within the field, with researchers actively working on methods to make AI algorithms more explainable and transparent.

Furthermore, the deployment of AI in investing raises ethical considerations. Algorithmic bias, for instance, can perpetuate and even amplify existing inequalities in the financial markets. AI systems trained on historical data might inadvertently reinforce past discriminatory practices, leading to unfair outcomes for certain groups of investors. Therefore, responsible development and deployment of AI in finance is paramount. This requires careful consideration of ethical implications at every stage, from data collection to model deployment, to ensure fairness and transparency. The potential for misuse, such as using AI to manipulate markets or exploit vulnerabilities, also necessitates robust regulatory oversight.

Another critical aspect is the integration of AI with human expertise. While AI can process data and identify patterns with incredible speed and accuracy, human judgment and intuition remain invaluable. Humans can assess qualitative factors, such as geopolitical events or shifts in consumer sentiment, that might be difficult for AI to quantify. The most successful investment strategies of the future will likely involve a synergistic blend of AI-driven analytics and human oversight, combining the strengths of both machine and human intelligence. This approach necessitates a skilled workforce capable of understanding and interpreting the insights generated by AI systems, while also maintaining a critical and discerning perspective.

In conclusion, AI is not replacing human investors; it's augmenting their capabilities. By providing access to advanced analytical tools and the ability to process massive datasets, AI is transforming the investment landscape. However, this transformation necessitates a nuanced understanding of AI's capabilities and limitations, a commitment to responsible development and ethical considerations, and a recognition of the enduring value of human judgment and experience. The future of successful investing lies in the effective integration of AI and human expertise, a powerful partnership poised to navigate the complexities of the modern financial world. The journey towards this future is one that requires continuous learning, adaptation, and a commitment to ethical and responsible innovation. The potential benefits are significant, but realizing them fully demands careful planning, rigorous oversight, and a balanced approach that appreciates the unique contributions of both human and artificial intelligence.

The transformative potential of AI in investment management extends beyond theoretical concepts and into the tangible realm of readily available tools and platforms. These range from sophisticated algorithmic trading systems to user-friendly robo-advisors, all designed to leverage the power of AI to enhance investment strategies. Understanding these tools and their capabilities is crucial for any investor seeking to navigate the evolving landscape of financial markets.

One of the most widely accessible applications of AI in investing is the rise of robo-advisors. These automated platforms utilize algorithms to create and manage investment portfolios based on individual investor profiles and risk tolerance. Robo-advisors often employ sophisticated asset allocation models, drawing on machine learning to optimize portfolio performance over time. They typically require minimal human intervention, offering a convenient and potentially cost-effective alternative to traditional financial advisors. However, it's crucial to understand the limitations. Robo-advisors often operate within pre-defined parameters, which may not adapt dynamically to unforeseen market events or individual investor circumstances. While their algorithms are designed to adapt, they may not react to events that fall outside the parameters of their programmed responses or learning models. Furthermore, the level of customization available may be limited compared to working with a human advisor who can provide more nuanced and personalized guidance. Before engaging with a robo-advisor, carefully review their investment strategies, fees, and potential risks. Consider whether their capabilities align with your financial goals and risk tolerance.

Moving beyond robo-advisors, we enter the realm of algorithmic trading platforms. These platforms utilize advanced algorithms to execute trades automatically, often at speeds and frequencies far exceeding human capabilities. These algorithms can be designed to identify and exploit fleeting market inefficiencies, react to news events in real-time, or implement complex hedging strategies. Algorithmic trading is heavily reliant on AI-powered systems analyzing vast quantities of data, including historical price patterns, market sentiment, and economic indicators. Sophisticated algorithms, often involving machine learning and deep learning, can identify complex correlations and predict future price movements with varying degrees of accuracy. However, the high-frequency nature of algorithmic trading introduces inherent risks. The speed at which these algorithms operate can amplify volatility and lead to flash crashes if not properly managed. Furthermore, the "black box" nature of some algorithms – the difficulty of understanding their inner workings – can be a significant concern. Transparency and explainability are paramount; investors should understand the underlying logic and parameters of any algorithm employed in their investments. The complexity and potential for unforeseen consequences highlight the importance of oversight and risk management when engaging with algorithmic trading platforms.

For investors seeking a more hands-on approach, several AI-powered analytical dashboards are available. These platforms provide sophisticated data visualization and analysis tools, allowing investors to monitor their portfolios, analyze market trends, and identify potential investment opportunities. Many dashboards incorporate machine learning algorithms to provide insights and predictions based on vast datasets. For example, these dashboards might offer sentiment analysis of social media conversations about specific companies, allowing investors to gauge public opinion and potential market reactions.

They might also highlight correlation between economic indicators and asset performance, allowing for more informed decision making. The accessibility of such information empowers investors to actively engage in their investment strategies, but it also requires a discerning eye. The visualizations and predictive capabilities of these dashboards are only as accurate as the data they are based on and the efficacy of the underlying algorithms. Investors must remain aware of potential biases in the data or limitations in the predictive capabilities of the system.

Beyond specific platforms, there's a growing ecosystem of AI tools tailored to various investment needs. AI-powered sentiment analysis tools, for instance, can track media coverage and social media discussions to gauge market sentiment toward specific stocks or industries. These tools can be valuable supplements to traditional fundamental and technical analysis, but they should not be the sole basis for investment decisions. Similarly, AI-driven risk management tools can assess portfolio risk exposure and provide recommendations for diversification or hedging. These sophisticated tools utilize machine learning algorithms to identify and quantify risks that might be overlooked by traditional methods. However, it's crucial to remember that AI is not a substitute for human judgment. It is a valuable tool that enhances decision-making, but not a guarantee of success.

While the benefits of AI-powered investment tools are significant, it's imperative to recognize the associated risks. Algorithmic bias, for instance, remains a significant concern. AI systems trained on historical data may perpetuate existing biases present in the market, leading to unfair or discriminatory outcomes. It's crucial for developers to actively address these biases and ensure fairness in algorithms used in investment tools. Additionally, the "black box" problem, the difficulty of understanding how some AI algorithms arrive at their conclusions, presents challenges for transparency and accountability. Regulatory oversight and increased emphasis on explainable AI are essential to mitigate these risks.

Moreover, security concerns are paramount. AI-powered platforms are potential targets for cyberattacks, and breaches could expose sensitive investor data or disrupt trading operations. Robust cybersecurity measures are therefore indispensable. The potential for misuse, such as using AI for market manipulation or insider trading, underscores the importance of ethical considerations and responsible development. The use of AI in finance must adhere to the highest ethical standards and be subject to strict regulatory oversight to maintain market integrity and investor protection.

Furthermore, the reliance on AI-driven tools shouldn't overshadow the critical role of human expertise. While AI can analyze massive datasets and identify patterns, human intuition, qualitative judgment, and an understanding of the broader economic and geopolitical context are still indispensable. The most successful investment strategies often involve a synergistic collaboration between human analysts and AI-powered tools, combining the strengths of both. This necessitates a skilled workforce capable of critically evaluating the outputs of AI systems and integrating them into a comprehensive investment strategy. It is the integration of human expertise and AI capabilities that offers the greatest potential for navigating the complexities of the financial markets.

In conclusion, AI is rapidly transforming the investment landscape, offering investors access to a range of powerful tools and platforms. From robo-advisors to algorithmic trading systems and analytical dashboards, AI enhances investment capabilities, providing access to previously unattainable insights and automating complex tasks. However, responsible use is paramount. Investors must be aware of the potential limitations, risks, and ethical considerations associated with AI-driven tools. The effective integration of AI and human expertise remains the key to unlocking the full potential of this technology while mitigating the associated risks. The future of successful investing lies in a balanced approach that leverages the power of AI while recognizing the enduring value of human judgment and experience.

Continuous learning, adaptation, and a commitment to ethical and responsible innovation will be essential for navigating this transformative period in the financial world.

The previous section explored the exciting possibilities and inherent risks of AI in investment management. However, the most sophisticated AI algorithm cannot compensate for a lack of discipline in one's investment strategy. The core principles of successful investing—patience, long-term vision, appropriate risk tolerance, and prudent diversification—remain paramount, regardless of the technological tools at our disposal. In fact, the availability of powerful AI tools can exacerbate the consequences of undisciplined investing, potentially leading to impulsive decisions and increased volatility within a portfolio.

Consider the allure of high-frequency trading algorithms. Their speed and complexity promise rapid returns, attracting investors seeking quick profits. But for those lacking a disciplined approach, the temptation to chase short-term gains can lead to impulsive trading, disregarding fundamental valuation and long-term growth potential. The allure of immediate gratification can override sound judgment, resulting in losses that outweigh any fleeting profits. A disciplined investor, on the other hand, will recognize that these algorithms are tools, not guarantees, and will maintain a strategic long-term focus, resisting the urge to react to every market fluctuation.

Patience is a cornerstone of disciplined investing. The markets are inherently cyclical, experiencing periods of growth and decline. A disciplined approach necessitates the ability to weather market downturns without panic selling. The temptation to sell during a market correction, driven by fear and short-term losses, is common. However, this often results in locking in losses and missing out on the subsequent recovery. Historical data consistently demonstrates that long-term investors who remain disciplined during market corrections tend to outperform those who react emotionally. AI tools can provide valuable insights into market trends, but they cannot predict the timing of market bottoms or tops with certainty. A disciplined investor will use this data to inform their strategy but will not allow short-term volatility to derail their long-term plan.

Risk tolerance is another critical aspect of a disciplined investment mindset. AI algorithms can help assess and manage risk, providing sophisticated risk profiles and diversification strategies. However, the algorithm's assessment must be aligned with the investor's individual circumstances, risk appetite, and financial goals. An investor with a low risk tolerance should not be swayed by AI-driven suggestions to invest heavily in high-risk assets simply because the algorithm predicts high potential returns. A disciplined approach requires a careful assessment of one's own comfort level with risk and the selection of an investment strategy that aligns with this assessment. This often involves a blend of different asset classes to achieve optimal diversification, minimizing exposure to any single risk factor. AI can help in diversification, but the ultimate decision rests on the investor's understanding of their own risk profile.

Diversification remains a crucial element, minimizing the impact of any single investment underperforming. AI can help identify correlations between assets and suggest optimal diversification strategies, but it's crucial to remember that complete diversification is a goal, not always perfectly achievable. Unexpected events can still impact even well-diversified portfolios. A disciplined investor recognizes this and continues to monitor and adjust their portfolio over time, based on a thorough understanding of their risk profile and financial goals. Regular rebalancing, based on pre-defined strategies, helps maintain the desired asset allocation, correcting any deviations caused by market fluctuations. This discipline ensures that the portfolio remains aligned with the investor's risk tolerance and long-term objectives.

Let's illustrate the importance of a disciplined approach with some real-life scenarios. Consider an investor who uses an AI-powered robo-advisor to manage their portfolio. The algorithm suggests a shift toward higher-risk assets due to a predicted market upturn. An undisciplined investor might impulsively follow this suggestion, increasing their risk exposure beyond their comfort level. However, if the predicted upturn fails to materialize, they may experience significant losses. A disciplined investor, on the other hand, would carefully evaluate the suggestion, considering their own risk tolerance and long-term goals before making any changes. They might partially adjust their portfolio, maintaining a balance between risk and reward.

Another scenario involves an investor using AI-powered sentiment analysis tools to gauge market sentiment toward a specific stock. An undisciplined investor might react immediately to a surge of positive sentiment, buying the stock at inflated prices. If the positive sentiment proves short-lived, they might find themselves holding a depreciating asset. A disciplined investor would analyze the sentiment data within the broader context of the company's fundamentals and the overall market conditions. They would consider the potential risks before making any investment decisions and would avoid impulsive trades based solely on short-term sentiment shifts.

These examples highlight how a disciplined approach, grounded in fundamental investment principles, is vital even in an AI-driven world. AI provides powerful tools to analyze data, identify patterns, and manage risk, but these tools should augment, not replace, the investor's judgment and discipline. The temptation to chase quick profits or react emotionally to market fluctuations must be resisted. Long-term planning, patience, appropriate risk tolerance, and diversification remain the cornerstones of successful investing, irrespective of the technological advancements.

Furthermore, a disciplined investor will regularly review and adjust their investment strategy, taking into account changes in their personal circumstances, market conditions, and the outputs of AI-powered tools. This requires a continuous learning process, staying updated on new technological advancements and their potential impact on the financial markets. It also necessitates a critical assessment of the information provided by AI tools, recognizing their limitations and potential biases. The ability to discern between actionable insights and noise is essential. A disciplined investor will not blindly trust the recommendations of any algorithm but will instead use them as a tool to enhance their own informed decision-making.

In essence, the adoption of AI in investment management is a double-edged sword. It empowers investors with powerful tools but simultaneously increases the potential for impulsive and ill-informed decisions. A disciplined investment mindset acts as the necessary counterbalance. It's the unwavering commitment to long-term goals, the ability to manage risk effectively, and the fortitude to withstand market volatility that ultimately determines the success of an investment strategy. The future of investing isn't solely about harnessing the power of AI, but about integrating it responsibly within a well-defined and disciplined investment framework. The most successful investors will be those who combine the power of technology with the timeless virtues of patience, discipline, and sound financial planning. The algorithms might change, the markets might fluctuate, but the fundamental principles of responsible investing will endure.

Building a robust investment strategy begins with a clear understanding of your financial goals. These goals provide the roadmap, guiding your investment decisions and ensuring that your actions align with your aspirations. Without well-defined goals, your investments become aimless, lacking a clear purpose and direction. This can lead to inconsistent decisions, emotional trading, and ultimately, suboptimal results.

Let's start by differentiating between short-term and long-term financial goals. Short-term goals typically involve objectives that you aim to achieve within a timeframe of one to five years. Examples include saving for a down payment on a house, funding a wedding, paying off high-interest debt, or covering unexpected expenses. These goals usually necessitate more conservative investment strategies, prioritizing capital preservation and liquidity over aggressive growth. You might consider savings accounts, money market funds, or short-term bonds to achieve these short-term goals, minimizing the risk of losing principal.

Long-term goals, on the other hand, extend beyond five years, often spanning decades. Common examples include retirement planning, funding children's education, or building a substantial investment portfolio. These goals typically allow for more aggressive investment strategies, leveraging the power of compounding and time to achieve significant growth. Investments such as stocks, mutual funds, and real estate may be suitable for achieving these long-term objectives.

The process of setting clear financial goals involves a detailed self-assessment. Consider your current financial situation, including your income, expenses, assets, and liabilities. Create a comprehensive budget to understand your cash flow and identify areas where you can allocate resources toward your investment goals. Be realistic in evaluating your financial capacity; avoid setting goals that are unattainable or overly ambitious. A balanced and achievable plan is far more effective than an overly optimistic, unattainable one.

Once you have a clear picture of your financial situation, start defining your specific goals. For each goal, specify the desired outcome, the timeframe for achieving it, and the estimated amount of capital required. For example, if your goal is to save for a down payment on a house in three years, specify the desired amount, research average house prices in your target area, and determine the necessary down payment percentage. Similarly, if your goal is to retire comfortably in 20 years, you'll need to project your retirement expenses and determine the required nest egg.

Determining your risk tolerance is another crucial step in formulating your investment strategy. Risk tolerance refers to your ability and willingness to accept the possibility of losing money in pursuit of higher returns. Your risk tolerance is influenced by various factors, including your age, investment horizon, financial situation, and personality. Younger investors with longer time horizons generally have a higher risk tolerance than older investors nearing retirement. Those with more financial resources can typically withstand greater losses without significantly impacting their overall financial security.

Several tools can help you assess your risk tolerance. Numerous online questionnaires and assessments are available, providing a personalized assessment of your risk profile. These questionnaires typically ask a series of questions about your investment experience, financial goals, and comfort levels with different investment types. The results of these assessments are often categorized into risk profiles, such as conservative, moderate, or aggressive, suggesting suitable investment options for each category.

However, you should always consider your individual circumstances beyond any standardized questionnaire. Remember that risk tolerance is not static; it can change over time depending on factors such as significant life events, changes in income, or market volatility. Regularly reassessing your risk tolerance and adapting your investment strategy accordingly is important to maintain alignment with your goals and comfort level.

Once you've defined your goals and assessed your risk tolerance, you can start translating them into a concrete investment strategy. This involves selecting specific investment vehicles that align with your risk profile and timeline. For short-term goals, prioritize investments with low risk and high liquidity, such as savings accounts or money market funds. For long-term goals, you can consider a broader range of options, including stocks, bonds, and real estate, balancing risk and return based on your individual risk tolerance.

Diversification plays a crucial role in mitigating risk. By spreading your investments across different asset classes, you reduce the impact of poor performance in any single investment. A well-diversified portfolio includes a mix of stocks, bonds, real estate, and potentially other alternative investments, balancing risk and return according to your goals and risk profile. However, diversification alone does not eliminate risk; unforeseen events can still affect even the most diversified portfolios. The goal is to reduce overall portfolio volatility, not eliminate all risk.

Regularly monitoring your investment portfolio's performance is vital. Track your investments' progress against your defined goals and adjust your strategy as needed. Market fluctuations, changes in your personal circumstances, or unexpected events may necessitate adjustments to your investment plan. Rebalancing your portfolio periodically, to maintain your desired asset allocation, helps ensure that you remain aligned with your risk tolerance and long-term goals. Remember that consistent monitoring and adjustments are essential to long-term success. It is not a "set it and forget it" endeavor.

Creating a personalized financial plan is crucial for achieving your financial objectives. A financial plan should not be seen as a static document; it should be a living, evolving document that adapts to your changing circumstances and goals. Regularly reviewing your financial plan, making necessary updates, and seeking professional advice when necessary are important aspects of long-term financial success. Many online resources and financial professionals can provide templates and guidance to help you build a comprehensive and personalized financial plan.

In conclusion, setting clear financial goals and developing a well-defined investment strategy are fundamental to achieving financial success. This process involves self-assessment, careful goal setting, risk tolerance evaluation, and the selection of suitable investment vehicles aligned with your objectives and risk profile. Consistent monitoring, regular adjustments, and the willingness to seek professional advice when necessary will further enhance the effectiveness of your plan. Remember, the journey towards your financial goals is a marathon, not a sprint. Discipline, consistency, and a well-defined plan are the cornerstones of long-term financial success. The integration of AI tools can further enhance the process, but it is not a replacement for thoughtful planning and disciplined execution.

Having established the groundwork for effective investment strategies, we now turn our attention to a powerful technological advancement that is reshaping the financial landscape: robo-advisors. These AI-powered platforms are transforming how individuals approach investment management, offering automated portfolio construction, rebalancing, and ongoing management at a fraction of the cost of traditional human advisors. Understanding their mechanics, capabilities, and limitations is crucial for anyone seeking to leverage technology to enhance their financial well-being.

At their core, robo-advisors utilize sophisticated algorithms and machine learning techniques to analyze an investor's financial profile, risk tolerance, and investment goals. This information, typically gathered through an online questionnaire, forms the basis for constructing a personalized investment portfolio. Unlike traditional advisors who may rely on subjective judgment and experience, robo-advisors rely on data-driven insights to make objective investment decisions. This objectivity minimizes the influence of emotional biases, a common pitfall in human investment behavior. The algorithms consider factors such as asset allocation, diversification, and tax optimization to create a portfolio designed to maximize returns while mitigating risk.

The process typically begins with a comprehensive assessment of the investor's financial situation. Questions regarding age, income, net worth, investment experience, and risk tolerance are common. Based on these responses, the robo-advisor generates a risk profile, categorizing the investor as conservative, moderate, aggressive, or somewhere in between. This risk profile dictates the asset allocation of the portfolio, influencing the proportion of stocks, bonds, and other assets included. For instance, a conservative investor might have a portfolio heavily weighted toward bonds and low-risk investments, while an aggressive investor might have a larger allocation to stocks and potentially more volatile asset classes.

Once the risk profile and asset allocation are determined, the robo-advisor automatically constructs the portfolio. This typically involves investing in exchange-traded funds (ETFs) or mutual funds, chosen based on their alignment with the investor's goals and risk tolerance. ETFs and mutual funds offer diversification, providing exposure to a wide range of stocks or bonds, reducing the risk associated with holding individual securities. The robo-advisor selects funds based on factors such as expense ratios, historical performance, and diversification within the specific asset class. The entire process is designed to be transparent and easily understandable by the investor. Regular reports provide a clear picture of portfolio performance, asset allocation, and any adjustments made.

A key advantage of robo-advisors is their ongoing portfolio rebalancing. Market fluctuations cause the proportions of assets in a portfolio to drift over time. Rebalancing involves adjusting the portfolio to maintain the original asset allocation determined by the risk profile. For instance, if a portfolio initially allocated 60% to stocks and 40% to bonds, and the stock market experiences significant growth, the proportion of stocks might increase beyond 60%. A robo-advisor automatically rebalances the portfolio by selling some stocks and buying bonds to restore the original 60/40 allocation. This process helps maintain the desired level of risk and ensures the portfolio remains aligned with the investor's goals.

Beyond portfolio construction and rebalancing, many robo-advisors offer additional features. Tax-loss harvesting, a strategy to minimize capital gains taxes, is becoming increasingly common. This involves selling losing investments to offset gains, reducing the overall tax burden. Some robo-advisors also offer financial planning tools, such as retirement calculators and budgeting assistance, providing a more comprehensive financial management solution. The level of sophistication and features varies significantly between different platforms, and understanding these differences is crucial for making an informed choice.

A critical consideration when choosing a robo-advisor is the fees charged. Fees typically vary depending on the assets under management (AUM), with higher balances often attracting lower fees. However, some robo-advisors charge flat fees or tiered fees based on the level of service provided. Transparency in fee structures is crucial, and investors should carefully review the fee schedule before making a commitment. Comparing fees across different robo-advisors is essential to finding the most cost-effective option. It's also important to understand what services are included in the fee. Some robo-advisors offer only basic portfolio management, while others provide more comprehensive financial planning and support.

Let's examine a few popular examples of robo-advisors to illustrate the diversity of offerings in the market. Betterment, for example, is known for its user-friendly interface and sophisticated tax-loss harvesting capabilities. Their tiered fee structure balances cost and service levels. Wealthfront, another established player, focuses on tax-efficient investing strategies and offers features such as direct indexing. Schwab Intelligent Portfolios, backed by a reputable brokerage, integrates seamlessly with other Schwab services, attracting investors already familiar with the platform. These are just a few examples, and numerous other robo-advisors cater to different investor needs and preferences. The ideal platform will depend on individual circumstances, risk tolerance, and investment goals.

While robo-advisors offer numerous advantages, it's essential to recognize their limitations. They are best suited for investors who are comfortable managing their investments primarily through an online platform. They generally don't offer the personalized attention and in-depth financial planning provided by traditional human advisors. Moreover, complex financial situations, such as significant inheritance or estate planning needs, often require the expertise of a human advisor. Therefore, robo-advisors are a powerful tool for many, but they are not a one-size-fits-all solution. They are most effective when viewed as a supplement to, rather than a complete replacement for, skilled financial guidance.

The rise of robo-advisors marks a significant shift in the financial services industry. By leveraging AI and data science, they democratize access to sophisticated investment management, making it more affordable and accessible to a broader range of investors. However, responsible usage necessitates a clear understanding of one's financial goals, risk tolerance, and the limitations of these technological solutions. Due diligence, careful comparison of platforms, and a realistic assessment of personal needs are crucial for successfully integrating robo-advisors into a comprehensive investment strategy. The future of personal finance is likely to involve a blend of technology-driven solutions and human expertise, offering a tailored approach to wealth management that fits the needs of a diverse investor base.

Selecting a robo-advisor requires a careful consideration of various factors, extending beyond simply comparing fees. The level of customization offered by the platform is paramount. Some robo-advisors offer highly personalized portfolios based on detailed questionnaires and assessments, while others offer more generic options. The investment philosophy of the platform is also crucial. Some may focus on passive investing, tracking market indexes, while others may employ more active strategies, attempting to outperform the market. Understanding these differences is critical to selecting a platform aligned with one's investment approach and risk tolerance.

The transparency of the platform's operations is another vital aspect. Investors should have clear access to information regarding portfolio composition, investment strategies, and fee structures. The platform should provide regular reports detailing portfolio performance and any adjustments made. Ease of use and customer service are equally important. A user-friendly interface can significantly enhance the investment experience, while responsive customer service can address any questions or concerns promptly and effectively.

Finally, the security and regulatory compliance of the platform are critical. Investors should ensure that the robo-advisor operates under appropriate regulatory oversight and employs robust security measures to protect their assets and personal information. Checking the platform's track record, security protocols, and regulatory approvals is a vital part of the due diligence process. Ultimately, the selection of a robo-advisor should be a well-informed decision, reflecting a thorough understanding of one's financial goals, risk tolerance, and the capabilities and limitations of the available technology. The integration of these powerful tools into a broader financial strategy requires careful planning and ongoing monitoring.

Algorithmic trading, often referred to as automated trading, high-frequency trading (HFT), or black-box trading, represents a significant advancement in the financial markets. It leverages sophisticated computer programs to execute trades based on pre-defined algorithms, eliminating the emotional biases and limitations of human decision-making. These algorithms analyze vast quantities of market data – including price movements, volume, order flow, and even news sentiment – to identify trading opportunities and execute trades at optimal times. The speed and efficiency of algorithmic trading are unparalleled, allowing for the execution of thousands, even millions, of trades in a single day.

The core of algorithmic trading lies in its ability to process and interpret massive datasets far beyond the capacity of a human trader. These algorithms can identify subtle patterns and correlations that might escape human observation, leading to potentially profitable trading opportunities. They are designed to react swiftly to market changes, capitalizing on short-term price fluctuations or exploiting arbitrage opportunities – instances where the same asset trades at different prices across various markets. This speed and precision are crucial in today's highly competitive trading environment.

Several factors contribute to the success of algorithmic trading strategies. Firstly, the use of backtesting – simulating the strategy on historical data – allows traders to assess the potential profitability and risk before deploying the algorithm in live markets. This reduces the risk of substantial losses, a crucial advantage over discretionary trading where decisions are made based on intuition or subjective judgment. Secondly, the algorithms are designed to adapt to changing market conditions. Many incorporate machine learning techniques, allowing them to learn from past performance and refine their strategies over time. This adaptability is essential for long-term success in the dynamic world of finance. Finally, strict risk management protocols are often integrated into algorithmic trading systems, limiting potential losses and protecting capital. Stop-loss orders, for instance, automatically exit a position when the price drops below a predetermined threshold, preventing significant losses.

However, algorithmic trading is not without its limitations and risks. One significant concern is the potential for errors in the code. A flaw in the algorithm's logic could lead to unintended consequences, resulting in substantial losses. Rigorous testing and validation are therefore crucial before deploying any algorithmic trading strategy. Another challenge lies in the unpredictable nature of the market. Even the most sophisticated algorithms cannot perfectly anticipate every market event, and unexpected shocks, such as geopolitical crises or sudden economic shifts, can significantly impact performance. The complexity of these algorithms also presents a barrier to entry for many smaller investors. Developing and maintaining sophisticated trading algorithms requires substantial technical expertise and financial resources, making it inaccessible to the average individual investor.

Several distinct types of algorithmic trading strategies exist, each with its own characteristics and risks. Mean reversion strategies, for example, capitalize on the tendency of prices to revert to their historical average. The algorithm identifies assets that have deviated significantly from their mean and places trades anticipating a return to the average. Momentum strategies, on the other hand, identify assets exhibiting strong upward or downward trends and bet on the continuation of this momentum. These strategies often involve high leverage and risk, as they are susceptible to reversals. Arbitrage strategies exploit price discrepancies between different markets for the same asset. For example, an algorithm might identify an instance where a stock trades at a slightly higher price on one exchange compared to another and execute trades to profit from this difference. However, these arbitrage opportunities are often short-lived and require extremely low latency trading infrastructure.

High-frequency trading (HFT) represents a specialized form of algorithmic trading characterized by extremely high speeds and volumes. HFT algorithms execute millions of trades per day, exploiting minute price differences and capturing small profits on a massive scale. This strategy relies heavily on sophisticated technology and infrastructure, including co-location servers placed directly within exchanges to minimize latency. While HFT can generate significant profits, it also carries significant risks, including the potential for market manipulation and the creation of unintended market instability. Regulatory scrutiny of HFT is therefore intense, aiming to mitigate these risks.

Statistical arbitrage is a more sophisticated algorithmic trading strategy that uses statistical models to identify mispricings in the market. These models consider a range of factors, including historical correlations between assets, economic indicators, and news sentiment, to predict future price movements. Unlike simpler strategies relying solely on price data, statistical arbitrage incorporates a broader range of information to improve prediction accuracy. However, it requires substantial expertise in statistical modeling and data analysis, along with the computational power to process large datasets.

The success of an algorithmic trading strategy depends on several factors, including the quality of the algorithm, the accuracy of its underlying models, and the efficiency of its execution. It also depends on market conditions and the overall investment environment. Strategies that perform well in one market environment might perform poorly in another. Therefore, thorough backtesting, risk management, and ongoing monitoring are crucial for managing the risks and maximizing the potential rewards.

The implementation of algorithmic trading involves several key steps. First, a robust trading strategy must be developed, based on rigorous research and analysis. This strategy is then translated into code, typically using programming languages such as Python or C++. The code is meticulously tested and validated using historical data, ensuring that it performs as expected. Once validated, the algorithm is deployed into a live trading environment, often integrated with a brokerage account or a dedicated trading platform. Continuous monitoring and performance analysis are essential to identify potential issues and adjust the strategy as market conditions change. Regular maintenance and updates are also necessary to ensure the algorithm continues to perform optimally.

Examples of successful algorithmic trading strategies abound, though specifics are generally kept confidential due to competitive pressures. Many hedge funds and institutional investors employ sophisticated algorithmic trading strategies, achieving impressive returns. However, numerous instances of unsuccessful strategies also exist, often due to unforeseen market events, algorithmic errors, or inadequate risk management. The failure of a single algorithmic trade can result in significant losses, highlighting the importance of rigorous testing and risk mitigation. The history of algorithmic trading is replete with both spectacular successes and dramatic failures, showcasing the double-edged sword of this technology.

In conclusion, algorithmic trading represents a powerful tool for investors, but it is not a panacea. Successful algorithmic trading requires expertise in computer science, finance, and statistics, as well as a deep understanding of market dynamics. It's crucial to remember that no algorithm can guarantee profits. The complexities, risks, and uncertainties of the financial markets necessitate a thoughtful and cautious approach, even when employing the most sophisticated technologies. The combination of cutting-edge technology and sound financial principles is essential for long-term success in the realm of algorithmic trading. The future of finance will likely witness an even greater integration of AI and algorithmic trading, further transforming the landscape of investment management and financial markets. However, this evolution will demand a continuous adaptation to new technologies and a vigilant approach to managing the inherent risks.

Building a personalized automated investment portfolio represents a significant leap forward in accessible and efficient wealth management. Gone are the days of solely relying on human advisors or navigating complex investment strategies alone. AI-powered tools, particularly robo-advisors, have democratized portfolio management, making sophisticated investment strategies available to a broader range of investors. This section will guide you through the process of constructing your own personalized automated investment portfolio, emphasizing the practical steps and considerations involved.

The first crucial step is defining your investment goals and risk tolerance. This involves a thorough self-assessment of your financial situation, future aspirations, and comfort level with potential losses. What are you saving for? Retirement? A down payment on a house? A child's education? The time horizon for your investments significantly influences your asset allocation. A longer time horizon allows for greater risk-taking, as there's more time to recover from potential market downturns. Conversely, shorter-term goals require a more conservative approach to minimize the risk of losses before the target date.

Your risk tolerance is equally important. Are you comfortable with the potential for significant fluctuations in your portfolio value? Or do you prefer a more stable, albeit potentially lower-return, investment strategy? Many robo-advisors incorporate questionnaires to assess your risk tolerance, guiding you toward appropriate asset allocations. These questionnaires typically delve into your investment experience, financial knowledge, and emotional response to market volatility. Honest self-reflection in this area is paramount; choosing an investment strategy that's too aggressive for your risk profile could lead to unnecessary stress and potentially substantial losses.

Once you've established your investment goals and risk tolerance, you can start selecting suitable assets. This typically involves a mix of asset classes, such as stocks, bonds, and potentially alternative investments like real estate or commodities. The proportion of each asset class in your portfolio is known as your asset allocation. A well-diversified portfolio spreads risk across various asset classes, mitigating the impact of poor performance in any single asset. For instance, a portfolio heavily weighted towards stocks might experience significant volatility, whereas a more balanced allocation that includes bonds can offer some stability during market downturns.

The allocation itself is heavily influenced by your risk profile and investment time horizon. Younger investors with longer time horizons might choose a portfolio with a higher allocation to stocks, aiming for potentially higher long-term returns. Conversely, older investors closer to retirement might opt for a more conservative allocation, prioritizing capital preservation over aggressive growth. Many robo-advisors offer pre-defined portfolios tailored to different risk profiles, simplifying the asset allocation process. However, you should understand the underlying asset allocation within these pre-defined portfolios to ensure they align with your individual circumstances and goals.

Several factors influence asset selection beyond your risk tolerance and time horizon. For example, your investment goals may lean towards specific sectors or industries. If you have a strong belief in the growth of renewable energy, you might favor stocks within that sector. Alternatively, you might choose index funds or exchange-traded funds (ETFs) that track specific market indices, providing broad market exposure without the need for individual stock picking. These passive investment strategies often have lower fees compared to actively managed funds.

Once your asset allocation and individual asset selection are determined, you need to allocate your funds. This is the stage where you actually invest your money in the chosen assets. Most robo-advisors handle this automatically, allocating your funds based on your chosen portfolio. However, if you're managing your portfolio manually, you'll need to execute the trades yourself, either through a brokerage account or directly with the platform offering the selected assets. Careful consideration should be given to transaction fees and commissions; higher fees can significantly eat into your returns over time.

Continuously monitoring your portfolio's performance is essential. While automated investment tools handle the day-to-day management, regular monitoring allows you to ensure the portfolio remains aligned with your goals. This involves reviewing your portfolio's performance periodically, assessing its overall returns and risk levels. Many robo-advisors offer convenient dashboards and reports showcasing key performance indicators, making monitoring simple and efficient. These tools often allow you to adjust your investment strategy based on your performance or changing circumstances. For example, if you've experienced significant gains, you might consider rebalancing your portfolio to realign with your original asset allocation. Conversely, if market conditions change, you may want to adjust your strategy to mitigate potential risks.

However, relying solely on automated tools isn't always sufficient. It's crucial to maintain a degree of human oversight. Periodically reviewing your investment goals and risk tolerance is essential, especially as your life circumstances change. Major life events, like marriage, the birth of a child, or changes in employment, could necessitate adjustments to your investment strategy. Regularly reviewing your portfolio's performance and ensuring it remains consistent with your updated goals is crucial for long-term success. This includes reviewing the underlying fees and expense ratios of the investment products you hold, as these can significantly impact your overall returns over time.

Beyond individual portfolio performance, staying informed about broader market trends and economic conditions is important. Understanding macroeconomic factors, such as interest rates, inflation, and geopolitical events, can help you make informed decisions regarding any necessary adjustments to your investment strategy. Although automated systems are designed to adapt to changing market conditions, a basic understanding of these factors can enhance your ability to make strategic decisions, ensuring that your long-term investment goals remain on track.

Let's illustrate this process with a hypothetical example. Imagine a 35-year-old individual with a long-term investment horizon, aiming to save for retirement. They have a moderate risk tolerance and want a diversified portfolio. Using a robo-advisor, they might select a pre-defined portfolio with a 70% allocation to stocks (including both US and international stocks) and a 30% allocation to bonds. The robo-advisor automatically allocates their funds according to this plan, regularly rebalancing the portfolio to maintain the target asset allocation. The individual regularly monitors the portfolio's performance through the robo-advisor's dashboard and makes adjustments as needed, perhaps reducing the equity exposure if market volatility increases significantly. This automated approach simplifies the investment process and allows the individual to focus on other aspects of their financial life while enjoying the benefits of a diversified, professionally managed portfolio.

In another scenario, a more risk-averse individual nearing retirement might choose a portfolio heavily weighted towards bonds, minimizing exposure to market fluctuations. They might opt for a less active strategy with less frequent rebalancing, prioritizing capital preservation. This exemplifies the versatility of personalized automated portfolios; the exact strategy is customized to meet the specific needs and goals of each individual investor. Regardless of the chosen strategy, regular monitoring and occasional adjustments are vital to ensure that the portfolio remains aligned with the investor's objectives.

To conclude, constructing a personalized automated investment portfolio empowers individuals to take control of their financial future. By carefully defining their goals, risk tolerance, and asset allocation, and by leveraging AI-powered tools to automate the management process, investors can enjoy significant benefits in terms of efficiency, accessibility, and professional-level portfolio management. Remember that while technology plays a crucial role, continuous monitoring, regular review, and a degree of informed human oversight are essential for long-term success and financial well-being. The integration of human judgment and technological assistance forms a powerful synergy in the pursuit of achieving financial objectives. The evolution of financial technology allows for both greater efficiency and a more personalized approach to wealth management, ultimately leading to a more accessible and effective system for all investors.

While the automation of investing offers significant advantages in terms of efficiency and accessibility, it's crucial to understand and manage the inherent risks. No investment strategy, however sophisticated, can guarantee profits, and automated systems are no exception. Effective risk management is paramount to protecting your capital and ensuring your investment goals are met. This involves a multi-pronged approach encompassing diversification, strategic order placement, and ongoing monitoring and adjustment.

Diversification remains a cornerstone of risk management in any investment portfolio, automated or otherwise. By spreading your investments across different asset classes – stocks, bonds, real estate, commodities – you reduce your dependence on the performance of any single asset. If one asset class underperforms, the others can potentially offset those losses, preventing a catastrophic drop in your overall portfolio value. The optimal level of diversification depends on your risk tolerance and investment goals. A younger investor with a longer time horizon might tolerate a higher allocation to equities, while an older investor nearing retirement might prefer a more conservative approach with a greater emphasis on fixed-income investments. Moreover, diversification extends beyond simple asset allocation. Within each asset class, you can further diversify geographically (e.g., investing in both US and international stocks) and across sectors (e.g., investing in technology, healthcare, and consumer staples).

Sophisticated AI-powered platforms often incorporate advanced diversification strategies, going beyond simple asset allocation. They may employ factor-based investing, focusing on specific characteristics like value, growth, or momentum, to construct portfolios that are less correlated with traditional market indices. This can lead to potentially higher risk-adjusted returns, but it's essential to understand the underlying strategies employed and their potential implications for risk. Many platforms offer transparency into their portfolio construction methodologies, allowing you to assess the level of diversification and understand the potential risks involved. Don't hesitate to explore these details; informed decision-making is crucial in managing risk effectively.

Beyond diversification, setting stop-loss orders is a crucial risk management technique. A stop-loss order is an instruction to your broker to sell a security when it reaches a specified price, limiting your potential losses. This is particularly useful in volatile markets, helping to protect your capital from significant declines. The placement of stop-loss orders is a strategic decision. Setting them too tightly can lead to premature exits from profitable positions, while placing them too loosely may not offer sufficient protection during market downturns. Consider your individual risk tolerance and the volatility of the specific assets in your portfolio when determining appropriate stop-loss levels. Many automated investing platforms offer tools and features that allow you to easily set and manage stop-loss orders, integrating this crucial risk management technique seamlessly into your automated investment strategy.

Regularly reviewing your portfolio's performance is another critical aspect of risk management. While AI-powered systems automate the trading process, they don't eliminate the need for human oversight. Consistent monitoring allows you to identify potential problems early and take corrective action. This includes assessing the overall performance of your portfolio in relation to your investment goals, as well as reviewing the performance of individual assets within your portfolio. Are your investments still aligned with your risk tolerance? Are any assets significantly underperforming compared to their peers? Are there any unforeseen external factors impacting your investment strategy? Automated platforms often provide detailed performance reports and dashboards, making portfolio monitoring straightforward. However, it's important to actively engage with these tools, going beyond simply observing the numbers. Actively analyzing the trends and patterns in your portfolio's performance is crucial for timely identification of potential risks.

Consider the impact of market events. A significant geopolitical event, for example, could negatively impact certain sectors or industries, prompting a review of your portfolio's exposure. Perhaps your portfolio has an overweight position in a particular sector that is suddenly facing headwinds. This requires a strategic reassessment and possible adjustments. You may need to rebalance your portfolio, shifting assets from underperforming areas to more promising ones. This rebalancing process, when done strategically, can help maintain your desired asset allocation and mitigate the impact of unforeseen market events.

Rebalancing itself is a dynamic risk management strategy. It involves selling some of your assets that have outperformed their targets and buying more of those that have underperformed. This ensures that your portfolio continues to adhere to its original asset allocation strategy, preventing any single asset from becoming overly concentrated and thereby increasing your overall portfolio risk. The frequency of rebalancing depends on various factors, including market conditions and your personal risk tolerance. Many robo-advisors offer automatic rebalancing features, making this process seamless and efficient. However, understanding the underlying logic behind automatic rebalancing is important, especially in more turbulent market conditions.

Risk management in automated investing is not a one-time event but an ongoing process. Regular reviews, strategic adjustments, and a proactive approach to monitoring market trends are essential for maintaining a healthy and well-performing portfolio. Leveraging the analytical capabilities of AI platforms can enhance your risk management strategies, but it's crucial to remain informed and actively engaged in the process.

Let's consider a practical example. An investor using an automated platform with a moderate risk profile might see their portfolio heavily weighted towards technology stocks. After a period of strong performance, technology stocks might represent a larger portion of their portfolio than originally intended. A regular portfolio review might reveal this concentration. The investor, using the platform's rebalancing feature, might then strategically sell some of the technology stocks and invest in bonds or other asset classes to realign their portfolio with their original risk profile. This timely action prevents overexposure to a single sector and mitigates potential losses in the event of a technology sector downturn.

Furthermore, external factors can significantly impact risk. Consider a sudden spike in inflation. This could negatively impact bond prices, a key component in many diversified portfolios. An investor aware of this macroeconomic trend might adjust their portfolio by reducing bond exposure and increasing holdings in assets that typically perform well during inflationary periods, such as real estate or commodities. The proactive response demonstrates the importance of staying informed about macroeconomic conditions and their impact on your investment strategy.

In conclusion, while AI and automation offer incredible benefits for managing investments, risk management remains a critical and ongoing responsibility for every investor. Diversification, stop-loss orders, and regular performance reviews are key tools that, when combined with a thorough understanding of market trends and economic conditions, allow for both the advantages of automation and the prudence of informed decision-making. The combination of technology and informed human intervention provides a robust framework for managing risk effectively and achieving long-term investment success. Remember that no automated system can completely eliminate risk; instead, the focus should be on mitigating it effectively. By employing a well-rounded approach that combines the strengths of automation with prudent, informed human oversight, investors can navigate the complexities of the market while maximizing their chances of achieving their financial goals.

Monitoring and adjusting your automated portfolio isn't a "set it and forget it" proposition. While AI handles the day-to-day trading, your active involvement in oversight and strategic adjustments is crucial for long-term success. Think of it as a partnership: AI provides the engine, and you provide the steering and navigation. This requires a proactive approach, incorporating regular reviews, a keen awareness of market shifts, and the ability to make informed adjustments.

One of the most valuable tools at your disposal is the comprehensive reporting typically provided by AI-powered investment platforms. These platforms often generate detailed performance reports, showing not just overall returns but also the performance of individual assets within your portfolio, their contribution to overall returns, and their risk profiles. These reports usually display your portfolio's performance relative to its benchmark, allowing you to gauge how well it's tracking against its objectives and identifying any significant deviations.

Regularly reviewing these reports—at least quarterly, if not monthly—is essential. Look beyond the headline numbers. Dive into the specifics. Are individual assets performing as expected? Are any assets significantly underperforming compared to their peers or benchmarks? Understanding the "why" behind performance is just as important as knowing the "what." A sudden dip in a specific stock might be due to company-specific news, while a broader market downturn could affect the entire portfolio. This level of granular analysis will help you identify potential problems early and take corrective action before they escalate.

Beyond individual asset performance, consider the overall allocation of your portfolio. Has the initial asset allocation, designed based on your risk tolerance and financial goals, drifted significantly? Perhaps a certain asset class, due to market fluctuations, now constitutes a larger portion of your portfolio than originally planned. This deviation might increase your risk exposure, making rebalancing a necessary step.

Market events significantly impact portfolio performance, necessitating constant vigilance. Geopolitical instability, economic downturns, or unexpected regulatory changes can dramatically affect asset prices. For example, a sudden surge in inflation could negatively impact bond prices, while rising interest rates might dampen the performance of growth stocks. Staying informed about such events and their potential implications is crucial for making well-timed adjustments. Follow reputable financial news sources, and consider subscribing to economic forecasts and analyses to stay ahead of potential market disruptions.

Rebalancing your portfolio is a crucial aspect of this ongoing monitoring process. It involves adjusting your asset allocation back to your predetermined target levels. If one asset class significantly outperforms others, rebalancing would entail selling a portion of those top performers and reinvesting in the underperforming ones. This ensures that your portfolio remains aligned with your initial risk tolerance and investment objectives, preventing excessive exposure to any single asset class.

The frequency of rebalancing depends on several factors, including your risk tolerance, investment goals, and market volatility. Some investors rebalance annually, while others do so more frequently (semiannually or even quarterly). AI-driven platforms often provide automated rebalancing tools, but understanding the underlying mechanics is vital. Make sure you understand the triggers that initiate rebalancing and the algorithms employed to make these adjustments. Don't blindly rely on automation; understand the "why" behind the adjustments recommended by your platform.

Let's consider a concrete example. Suppose an investor with a moderately conservative portfolio, initially allocated 60% to stocks and 40% to bonds, sees their stock holdings increase to 70% due to strong market performance. A regular portfolio review reveals this imbalance. To rebalance, the investor would sell a portion of their outperforming stocks, reinvesting the proceeds in bonds to restore the 60/40 ratio. This prevents overexposure to the stock market and reduces overall portfolio risk, aligning it with their original investment strategy.

Another example involves adapting to changing macroeconomic conditions. Imagine a scenario where rising inflation becomes a major concern. A savvy investor might review their portfolio and realize that their significant holdings in fixed-income securities, such as bonds, are particularly vulnerable. To mitigate this risk, they may adjust their portfolio by reducing bond exposure and shifting funds into assets generally considered inflation hedges, such as real estate or commodities. This proactive adjustment demonstrates the importance of reacting to changing economic circumstances.

Beyond rebalancing, consider incorporating other risk management strategies. Stop-loss orders, as discussed earlier, remain a valuable tool to protect your investments from significant losses. Regularly review the stop-loss levels for individual assets, adjusting them as needed based on market volatility and your risk tolerance. This ensures that your protection levels are appropriate and adapt to changing market conditions.

Technology plays a significant role in simplifying the monitoring and adjustment process. AI-powered platforms often provide user-friendly dashboards and visual tools to track portfolio performance, identify potential problems, and execute adjustments. These tools streamline the process, making it easier to stay informed and make timely decisions. However, don't let technology replace active engagement. Always understand the underlying logic behind any recommendations or adjustments suggested by your AI platform.

In conclusion, while AI automates the trading aspect of investing, it doesn't eliminate the need for your active participation in monitoring and adjusting your portfolio. Regular reviews, proactive responses to market changes, and strategic rebalancing are vital for ensuring your portfolio remains aligned with your investment goals and risk tolerance. Combine the efficiency of AI-driven automation with the informed decision-making of an active investor to navigate the complexities of the financial markets and maximize your chances of long-term success. Remember, the goal isn't to eliminate risk entirely—that's impossible—but to effectively manage and mitigate it, consistently working towards your financial objectives. This ongoing process, a blend of technological prowess and astute human oversight, is the key to unlocking the full potential of AI-powered investing.

Sentiment analysis represents a powerful advancement in leveraging AI for enhanced decision-making within the financial markets. It moves beyond the traditional reliance on numerical data, incorporating the qualitative aspect of market sentiment derived from vast quantities of textual information. This shift allows investors to gain a more nuanced and potentially predictive understanding of market trends, complementing the quantitative analysis already prevalent in modern finance.

The core of sentiment analysis lies in its ability to interpret the emotional tone expressed in textual data. This data can range from news articles and financial reports to social media posts, blogs, and online forums. By processing this information, sophisticated algorithms can identify the prevalent sentiment—positive, negative, or neutral—and even quantify the intensity of that sentiment. This ability to gauge the collective mood of the market offers a unique perspective that can significantly inform investment strategies.

Several techniques underpin sentiment analysis. One common approach involves lexicon-based methods, which rely on pre-defined dictionaries containing words and phrases associated with specific sentiments. These dictionaries are used to score the sentiment of a given text by tallying the number of positive and negative words present. While relatively straightforward, this method can be limited by its reliance on pre-defined lexicons, which might not always capture the nuances of language and the context in which words are used.

More advanced techniques employ machine learning algorithms, often trained on large datasets of labeled text. These algorithms can learn complex patterns and relationships between words and phrases, allowing them to accurately assess sentiment even in ambiguous or complex contexts. These machine learning models, especially deep learning models like recurrent neural networks (RNNs) and transformers, have proven exceptionally adept at understanding the subtleties of human language, significantly improving the accuracy and reliability of sentiment analysis.

The application of sentiment analysis in financial markets is multifaceted. For example, analysts can use it to gauge investor confidence in specific companies or industries. By analyzing news articles and social media chatter related to a particular company, analysts can identify shifts in public opinion that might precede changes in stock price. A surge of negative sentiment surrounding a company, for instance, could signal an impending drop in its stock price, providing investors with a valuable early warning signal. Conversely, a sudden upswing in positive sentiment might suggest a potential investment opportunity.

Furthermore, sentiment analysis can be used to assess the overall market sentiment. By analyzing a broad range of news sources and social media data, analysts can determine whether the overall market mood is optimistic or pessimistic. This information can be extremely helpful in making tactical asset allocation decisions. For instance, if the overall market sentiment turns significantly negative, an investor might consider reducing their exposure to equities and increasing their allocation to safer assets like bonds or cash.

The power of sentiment analysis extends beyond simply identifying positive or negative sentiment. It can also provide insights into the intensity and strength of that sentiment. Knowing that a particular news article expresses strong negative sentiment regarding a company, for example, provides a more compelling signal than simply knowing that the sentiment is negative. This granular level of information allows investors to make more informed decisions and potentially identify potentially high-impact events.

However, it's crucial to acknowledge the limitations of sentiment analysis. While it provides valuable insights, it's not a foolproof predictor of market movements. Market sentiment is just one factor among many that influence stock prices and overall market trends. Other factors, such as macroeconomic conditions, company-specific news, and regulatory changes, play a significant role and must be considered in conjunction with sentiment analysis.

Another limitation is the potential for noise and bias in the data. Social media, for instance, can be susceptible to misinformation and manipulation. Furthermore, the language used in different sources can vary significantly, potentially affecting the accuracy of sentiment analysis. Sophisticated algorithms can help mitigate these issues, but it's critical to use a combination of data sources and validation techniques.

Despite these limitations, sentiment analysis remains a powerful tool in the arsenal of the modern financial professional. When used responsibly and in conjunction with other analytical techniques, it can significantly enhance investment decision-making. A balanced approach, integrating sentiment analysis with traditional fundamental and technical analysis, provides a more comprehensive understanding of market dynamics, fostering more robust and well-informed investment strategies.

To illustrate its practical application, consider the following scenario: Imagine a major technology company is about to release its quarterly earnings report. In the days leading up to the report, sentiment analysis tools monitoring news articles and social media discussions reveal a growing wave of negative sentiment. Investors are expressing concerns about slowing growth, increasing competition, and potential regulatory challenges. This negative sentiment, if corroborated by other indicators, might lead a prudent investor to consider reducing their exposure to the company's stock before the earnings report is released, mitigating potential losses.

Conversely, imagine a promising biotech company announces a breakthrough in a new drug development. Sentiment analysis of news reports and online forums shows an immediate surge of overwhelmingly positive sentiment. This positive sentiment, coupled with fundamental analysis confirming the significance of the breakthrough, could prompt an investor to consider adding the company's stock to their portfolio. This illustrates how sentiment analysis can provide valuable early insights into market-moving events.

In conclusion, sentiment analysis represents a significant advancement in the application of AI to financial markets. By analyzing textual data, it captures the often overlooked qualitative aspects of market dynamics, offering investors a unique and valuable perspective. However, its success hinges on a careful and considered approach, combining it with traditional financial analysis techniques and acknowledging its inherent limitations. The future of financial decision-making will likely involve an increasingly sophisticated integration of AI-powered tools, including sentiment analysis, to navigate the complexities of the modern financial landscape. By leveraging these tools effectively and responsibly, investors can enhance their ability to make more informed and profitable investment decisions. The key lies in understanding not just what the algorithms are telling you, but also the "why" behind their analysis, allowing for a more nuanced and informed interpretation of market sentiment.

Predictive analytics represents a significant leap forward in leveraging AI for enhanced financial decision-making. Unlike sentiment analysis, which focuses on interpreting qualitative data, predictive analytics delves into the quantitative realm, utilizing historical data to forecast future market trends. This involves sophisticated algorithms that identify patterns and relationships within vast datasets, enabling predictions about everything from stock prices and interest rates to currency exchange rates and credit risk.

The foundation of predictive analytics in finance rests on the availability of large, high-quality datasets. These datasets typically encompass years, even decades, of historical market data, including price movements, trading volumes, economic indicators, and company financials. The richer and more comprehensive the dataset, the more accurate and reliable the predictions are likely to be. The process typically begins with data cleaning and preparation, ensuring the data's accuracy and consistency. Missing values need to be handled, outliers identified and addressed, and the data often needs transformation to be suitable for the chosen analytical techniques.

Several powerful techniques underpin predictive analytics. One common approach involves time series analysis, which focuses on identifying patterns and trends within sequential data points over time. This is particularly useful for forecasting stock prices, interest rates, and other financial variables that exhibit temporal dependencies. Autoregressive integrated moving average (ARIMA) models and exponential smoothing are frequently used time series methods. ARIMA models capture the autocorrelations within the time series data, while exponential smoothing assigns exponentially decreasing weights to older data points. The choice of model depends on the specific characteristics of the time series and the prediction horizon.

Another powerful technique involves machine learning algorithms, such as linear regression, support vector machines (SVMs), and neural networks. These algorithms can identify complex relationships between multiple variables, providing more nuanced predictions than simpler time series models. Linear regression models seek to establish a linear relationship between predictor variables and the target variable. SVMs find the optimal hyperplane to separate data points into different classes, making them effective for classification tasks like predicting whether a stock will go up or down. Neural networks, especially deep learning models, can capture non-linear relationships and are particularly adept at handling high-dimensional datasets with numerous predictor variables.

The selection of appropriate algorithms and features is crucial for the accuracy of predictive models. Feature engineering, the process of creating new variables from existing ones, can significantly improve model performance. For example, technical indicators derived from historical price data, such as moving averages, relative strength index (RSI), and Bollinger Bands, can serve as valuable predictor variables. Similarly, fundamental data, such as earnings per share (EPS), price-to-earnings ratio (P/E), and debt-to-equity ratio, can be incorporated to enhance predictive accuracy.

However, it is essential to acknowledge the limitations of predictive analytics. Financial markets are inherently complex and unpredictable, influenced by a multitude of factors that are not always easily quantifiable or predictable. Therefore, no predictive model can guarantee perfect accuracy. Moreover, the effectiveness of any model depends heavily on the quality and relevance of the data used to train it. Using outdated or biased data can lead to inaccurate and misleading predictions.

Another significant challenge is model overfitting. This occurs when a model is too complex and fits the training data too closely, resulting in poor performance on new, unseen data. To mitigate overfitting, techniques such as cross-validation and regularization are employed. Cross-validation involves splitting the data into multiple subsets, training the model on some subsets and testing it on others. Regularization adds penalties to the model's complexity, discouraging overfitting.

Furthermore, the accuracy of predictive models can deteriorate over time as market conditions change. Models trained on historical data might not perform well in significantly different market environments. Regular model retraining and recalibration are essential to maintain accuracy. This often requires monitoring the model's performance on new data and making adjustments as needed.

The interpretation of predictive analytics results also demands caution. The output of a predictive model should not be interpreted as a definitive forecast but rather as a probabilistic estimate. It's crucial to consider the model's confidence intervals and understand the range of possible outcomes. Over-reliance on predictive models without considering other factors, such as qualitative information and expert judgment, can lead to flawed decision-making.

Let's consider some specific examples. Predictive models have been successfully used to forecast credit risk, identifying individuals or businesses with a high probability of defaulting on loans. This allows financial institutions to make more informed lending decisions, minimizing potential losses. However, these models have limitations; they can't perfectly predict all defaults and can inadvertently discriminate against specific groups if not carefully designed and implemented.

Another area where predictive analytics has shown promise is algorithmic trading, where computer algorithms automatically execute trades based on predictive models. High-frequency trading (HFT) firms, for example, utilize sophisticated algorithms to exploit tiny price discrepancies in milliseconds. However, the success of algorithmic trading is contingent on the accuracy of the underlying predictive models and can be significantly impacted by unexpected market events or changes in trading regulations.

Conversely, there have been instances where predictive models failed to anticipate major market events. The 2008 financial crisis serves as a stark reminder of the limitations of predictive analytics. Many models failed to accurately predict the severity of the crisis due to unforeseen interactions among multiple factors and the inherent limitations of traditional statistical models.

Despite their limitations, predictive analytics remain a powerful tool in the financial industry. When used judiciously and in conjunction with other analytical techniques and expert judgment, they can greatly enhance decision-making. The key lies in a balanced approach, acknowledging the models' limitations, continuously monitoring performance, and adapting as needed. The future of finance will undoubtedly see an even greater integration of predictive analytics, with increasing sophistication in model development and data analysis. The continued advancement of AI and machine learning promises even more powerful and nuanced predictive models, but responsible implementation and careful interpretation will remain paramount.

Building upon the predictive power of AI in forecasting market trends, we now delve into the realm of AI-powered portfolio optimization. This represents a significant advancement in personalized investment management, moving beyond generic asset allocation strategies to create portfolios tailored to individual investor needs and risk profiles. The core principle is leveraging AI's ability to process vast datasets and identify optimal asset allocations that maximize returns while minimizing risk, all while adhering to predefined constraints.

Traditional portfolio optimization methods, like Markowitz Mean-Variance Optimization, rely on historical data to estimate expected returns and covariance matrices. While useful, these techniques have limitations. They often assume normally distributed returns, a simplification that doesn't always hold true in real-world markets. Furthermore, they are sensitive to estimation errors in input parameters, potentially leading to suboptimal portfolios. AI, particularly machine learning, offers a more robust and adaptable approach.

Machine learning algorithms excel at identifying non-linear relationships and complex patterns within data, surpassing the limitations of traditional statistical methods. Techniques like reinforcement learning (RL) have emerged as particularly promising for portfolio optimization. In RL, an AI agent learns to make optimal investment decisions through trial and error within a simulated market environment. The agent receives rewards for profitable trades and penalties for losses, continuously refining its strategy to maximize cumulative rewards over time. This approach effectively adapts to changing market conditions and can identify complex, non-linear trading strategies that might be overlooked by traditional methods.

Several variations of RL algorithms are employed in portfolio optimization. Deep Q-networks (DQNs) are a popular choice, using deep neural networks to approximate the optimal action-value function. This allows the agent to learn complex strategies and handle high-dimensional data representing diverse assets and market conditions. Proximal policy optimization (PPO) is another effective method, offering greater stability and efficiency in training compared to some other RL algorithms. It focuses on improving the policy gradually, reducing the risk of drastic performance fluctuations during learning.

Beyond RL, other machine learning algorithms contribute to AI-powered portfolio optimization. Genetic algorithms, inspired by natural selection, are employed to evolve optimal portfolios over generations. Each generation comprises a population of portfolios, with the fittest (those yielding the highest returns) selected for reproduction and mutation. This iterative process continues until a satisfactory portfolio is obtained. Support Vector Machines (SVMs) and other classification algorithms can be integrated to predict asset price movements, guiding the optimization process with more informed forecasts.

The implementation of AI-powered portfolio optimization involves several crucial steps. First, the definition of the investor's risk tolerance and financial goals is paramount. This might involve questionnaires, interviews, or analysis of past investment behavior. This information forms the crucial constraints within the optimization process, ensuring the generated portfolios align with individual needs. Second, the selection and preparation of data are critical. This includes historical price data for relevant assets, economic indicators, and other relevant factors such as sentiment data derived from news and social media. Data cleaning, handling missing values, and feature engineering are crucial steps to ensure the quality of the input data.

Next, the choice of AI algorithm depends on several factors including the complexity of the problem, available computational resources, and desired level of explainability. RL is often preferred for its ability to learn complex strategies, but it can be computationally intensive. Genetic algorithms provide a good balance between performance and computational efficiency. The selection process requires careful consideration of the trade-off between computational cost and optimization performance.

Once the algorithm is chosen, the training phase begins. This typically involves simulating market scenarios using historical or synthetic data, allowing the AI agent to learn optimal investment strategies. Backtesting on historical data provides an initial assessment of performance, but out-of-sample testing is crucial to evaluate generalization capabilities. The performance metrics to be optimized, such as Sharpe ratio, Sortino ratio, or maximum drawdown, must be clearly defined and aligned with the investor's risk profile.

Finally, the optimized portfolio is generated, providing a recommended allocation of assets based on the chosen AI model and defined constraints. The generated portfolio is not a static entity but should be dynamically rebalanced periodically as market conditions change. This involves retraining the AI model or adjusting parameters to adapt to new data and insights. Regular monitoring and re-optimization are crucial to maintain the portfolio's alignment with the investor's objectives.

The advantages of AI-powered portfolio optimization are numerous. It allows for the creation of highly personalized portfolios tailored to individual risk profiles and goals, enhancing the investor experience. AI algorithms can handle large datasets and complex relationships beyond the capacity of human analysts, potentially leading to superior investment outcomes. The ability to adapt to changing market conditions improves the resilience and robustness of the portfolio. Furthermore, AI can incorporate various data sources, including alternative data, providing a richer and more comprehensive view of the market.

However, it's crucial to acknowledge the limitations. The accuracy of AI-powered optimization depends heavily on the quality of the data used for training. Biased or incomplete data can lead to suboptimal or even harmful portfolios. Overfitting, where the model performs well on training data but poorly on new data, is a significant concern. Therefore, robust techniques such as cross-validation and regularization must be employed. Moreover, the "black box" nature of some AI algorithms can make it difficult to interpret the reasoning behind investment decisions, raising concerns about transparency and trust. The explainability of the model's decision-making process is crucial for investor confidence. Further research is focusing on developing more transparent and interpretable AI algorithms.

Specific examples illustrate the power of AI-driven portfolio optimization. Robo-advisors are already utilizing AI to offer automated portfolio management services to a broader range of investors. These platforms use algorithms to assess risk tolerance, generate diversified portfolios, and rebalance them regularly, offering a cost-effective and accessible investment solution. Hedge funds and large institutional investors are also increasingly leveraging AI to enhance their investment strategies. They use AI to identify subtle market trends, execute trades with greater speed and precision, and manage risks more effectively.

The development of AI-powered portfolio optimization is an ongoing process. Research is continuously advancing, leading to more sophisticated algorithms and improved techniques. The integration of alternative data, such as satellite imagery or social media sentiment, holds the potential to further enhance predictive capabilities and optimize portfolio performance. However, the responsible and ethical implementation of AI in finance remains paramount. Transparency, explainability, and rigorous testing are essential to ensure that AI-powered portfolio optimization benefits investors while mitigating potential risks. The future of investment management undoubtedly lies in the integration of AI, but careful consideration of the associated limitations and ethical implications remains paramount for the responsible and sustainable growth of this technology.

Building on the previous discussion of AI-powered portfolio optimization, we now turn our attention to the critical role of AI in analyzing market data. The sheer volume and velocity of financial data available today – encompassing stock prices, economic indicators, news sentiment, social media chatter, and alternative data sources like satellite imagery – far surpasses human capacity for analysis. AI offers a powerful solution, enabling the extraction of meaningful insights and patterns that would otherwise remain hidden.

The application of AI in market data analysis leverages several key techniques from the broader field of data science. At the foundation lies data preprocessing and cleaning. Raw market data is often incomplete, inconsistent, and noisy. AI algorithms require meticulously cleaned and prepared data to function effectively. This involves handling missing values, addressing outliers, and transforming data into suitable formats for algorithmic processing. Techniques like imputation for missing values, normalization or standardization for feature scaling, and outlier detection using methods such as IQR or DBSCAN are crucial preparatory steps.

Once the data is prepared, various AI techniques can be employed for analysis. Machine learning (ML) algorithms are particularly well-suited for uncovering patterns and relationships within complex datasets. Supervised learning, for example, can be used to predict future price movements based on historical data. This involves training a model on a labeled dataset, where the input features are market indicators (e.g., volume, moving averages, technical indicators) and the output is the future price change. Popular algorithms for this task include linear regression, support vector machines (SVMs), random forests, and gradient boosting machines (GBMs). The choice of algorithm depends on factors such as the dataset size, the complexity of the relationships between features and target, and the desired level of interpretability.

Deep learning (DL), a subfield of ML involving artificial neural networks with multiple layers, offers even greater power for analyzing complex data. Recurrent neural networks (RNNs), particularly Long Short-Term Memory (LSTM) networks, are adept at processing sequential data like time series of stock prices. LSTMs excel at capturing long-term dependencies in the data, which is crucial for accurate price forecasting. Convolutional neural networks (CNNs) can be applied to image data, such as satellite imagery that may reveal economic activity or infrastructure changes, providing valuable alternative data inputs for market analysis.

Beyond prediction, AI can be used for classification tasks in market analysis. For example, AI algorithms can classify companies based on their financial health, predict credit risk, or identify potential fraudulent activities. These classifications can guide investment decisions, risk management strategies, and regulatory compliance efforts. Unsupervised learning techniques, such as clustering and dimensionality reduction, also play an important role. Clustering algorithms, like k-means or hierarchical clustering, can group similar assets or market segments, identifying potential investment opportunities or risks. Dimensionality reduction techniques, such as Principal Component Analysis (PCA) or t-distributed Stochastic Neighbor Embedding (t-SNE), help to visualize and interpret high-dimensional data, revealing underlying structures and patterns that might be overlooked in raw data.

The application of AI in market data analysis extends beyond individual assets and into macroeconomic forecasting. AI algorithms can analyze vast amounts of economic data—inflation rates, unemployment figures, interest rates, GDP growth—to build predictive models for macroeconomic trends. These models can inform investment strategies, helping investors anticipate market shifts and adjust their portfolios accordingly. For instance, AI models can predict potential recessions by analyzing various economic indicators, providing valuable early warning signals to investors and policymakers. Moreover, AI can analyze news articles, social media posts, and other textual data to gauge market sentiment, providing further insights into potential market shifts. Natural Language Processing (NLP) techniques are instrumental in this process, enabling the analysis of vast quantities of unstructured textual data to identify prevailing opinions and expectations.

Furthermore, AI-driven sentiment analysis can provide a valuable layer of information that supplements traditional quantitative data. By analyzing news articles, social media posts, and other online sources, AI algorithms can gauge the overall market sentiment towards specific stocks or sectors. This information, while subjective, can be a valuable predictor of future price movements. For example, a surge in negative sentiment surrounding a particular company could signal an impending price drop, allowing investors to make informed decisions. The combination of quantitative and qualitative data, made possible by AI, leads to a much richer and more comprehensive understanding of the market dynamics.

However, the effective use of AI in market data analysis is not without its challenges. One key concern is the potential for bias in the data. If the training data reflects existing biases, the AI model will likely perpetuate and even amplify those biases, leading to inaccurate or unfair predictions. Careful data selection and rigorous bias mitigation techniques are crucial to ensure the fairness and reliability of the results. Another challenge is the interpretation of AI models' outputs. Many advanced AI algorithms, particularly deep learning models, are often referred to as "black boxes" due to their complex internal workings. Understanding why an AI model makes a particular prediction is essential for building trust and ensuring responsible use. Techniques such as explainable AI (XAI) aim to address this challenge by providing more transparent and interpretable insights into the decision-making process.

Despite these challenges, the potential benefits of AI in market data analysis are immense. AI systems can process vast amounts of data far more quickly and efficiently than humans, identifying subtle patterns and relationships that might be missed by human analysts. They can also adapt to changing market conditions more readily than traditional statistical models, adjusting their predictions and strategies in response to new information. The increasing availability of alternative data sources, such as satellite imagery or transaction data, further enhances the capabilities of AI in market analysis, unlocking opportunities for more informed and precise insights.

Real-world examples illustrate the power of AI in market data analysis. Many hedge funds and investment banks are already employing sophisticated AI algorithms to analyze market data, identify trading opportunities, and manage risk. These algorithms can sift through massive datasets, identifying subtle patterns and anomalies that may indicate potential market movements. Similarly, robo-advisors leverage AI to analyze client data and recommend suitable investment portfolios, providing personalized investment advice at a scale that would be impossible with traditional methods. Moreover, regulatory bodies are increasingly utilizing AI to detect and prevent fraudulent activities in the financial markets. AI algorithms can analyze transaction data and identify unusual patterns that might indicate money laundering, insider trading, or other forms of financial crime. The application of AI in market surveillance is constantly evolving, becoming more sophisticated and effective in protecting the integrity of financial markets.

In conclusion, AI is transforming the landscape of market data analysis, offering powerful tools to analyze vast amounts of data, identify complex patterns, and make more accurate predictions. However, responsible use requires careful attention to data quality, bias mitigation, and model interpretability. As AI technologies continue to advance, their impact on market analysis will only intensify, shaping the future of investment strategies, risk management, and regulatory oversight in the financial sector. The continuous development and refinement of AI algorithms, coupled with the increasing availability of high-quality data, promise even more sophisticated and effective market analysis in the years to come. The key lies in harnessing the power of AI responsibly, acknowledging its limitations, and ensuring its application aligns with ethical considerations and the principles of fairness and transparency.

Building upon our exploration of AI's role in market data analysis, we now arrive at the crucial juncture of translating these insights into actionable investment strategies. While AI offers unprecedented analytical power, its effectiveness hinges on its integration with established investment principles and a robust risk management framework. It's not a replacement for human judgment but rather a powerful augmentation, enhancing our ability to make informed decisions.

The process begins with a clear understanding of one's investment objectives and risk tolerance. AI can analyze vast datasets to identify potential investment opportunities aligned with these parameters, but the ultimate decision rests with the investor. For instance, an AI model might identify a particular stock as undervalued based on its analysis of financial statements, market sentiment, and other relevant data. However, the investor must independently assess the company's fundamental strength, competitive landscape, and potential risks before committing capital. AI can quantify the probability of success, but it cannot eliminate the inherent uncertainty associated with any investment.

A critical aspect of integrating AI into investment decisions is the judicious selection of AI tools and algorithms. The market offers a plethora of AI-powered platforms, each with its strengths and limitations. Some platforms specialize in quantitative analysis, utilizing sophisticated algorithms to identify statistical anomalies and predict market movements. Others focus on qualitative analysis, harnessing natural language processing (NLP) to gauge market sentiment and extract insights from news articles and social media. The choice of platform depends on the investor's specific needs and resources. A thorough understanding of the underlying methodologies employed by the AI tool is crucial to ensure its suitability and reliability.

Moreover, the output generated by AI models should always be treated with critical evaluation, not blindly accepted as gospel truth. AI algorithms are trained on historical data, which may not accurately reflect future market conditions. Unforeseen events, such as geopolitical upheavals or unexpected regulatory changes, can significantly impact market dynamics, rendering even the most sophisticated AI predictions inaccurate. Therefore, a healthy skepticism and continuous monitoring of the investment are essential. Regular reviews of the AI-generated insights, combined with updated fundamental analysis, help to maintain a flexible and adaptive investment strategy.

Effective risk management is paramount when incorporating AI into investment decision-making. AI can assist in identifying and quantifying various risks, but it cannot eliminate them entirely. Diversification remains a crucial risk mitigation strategy, and AI can aid in constructing diversified portfolios optimized for specific risk-return profiles. However, the investor must be aware of the limitations of AI-based risk assessment. AI models may not fully capture the complexities of interconnected risks, and unforeseen black swan events can still significantly impact even well-diversified portfolios. A robust risk management framework should involve regular portfolio reviews, stress testing, and contingency planning, incorporating both AI insights and traditional risk management techniques.

Furthermore, the ethical considerations surrounding the use of AI in investing should not be overlooked. The training data used to develop AI models can contain biases, potentially leading to unfair or discriminatory outcomes. It is essential to ensure that the AI tools employed are free from bias and that their outputs are transparent and interpretable. This requires careful scrutiny of the data used to train the AI models and a thorough understanding of how the AI algorithms arrive at their conclusions. Transparency in the decision-making process is essential for building trust and maintaining investor confidence.

Let's illustrate the practical application of AI insights with real-world examples. Consider a long-term investor interested in sustainable investments. An AI-powered platform specializing in Environmental, Social, and Governance (ESG) investing could analyze a vast dataset of company disclosures, news articles, and other relevant information to identify companies with strong ESG profiles. The platform might rank these companies based on various ESG metrics, helping the investor make informed decisions aligned with their ethical and sustainability goals. However, the investor should not solely rely on the AI's ranking but also conduct independent research to validate the platform's findings.

Another example involves a short-term trader seeking to capitalize on short-term market volatility. An AI-driven trading bot could analyze real-time market data, including price movements, volume, and sentiment, to identify short-term trading opportunities. The bot might use sophisticated algorithms to execute trades automatically, based on pre-defined parameters. This approach can significantly enhance trading speed and efficiency, but it also carries a higher level of risk. The investor must carefully monitor the bot's performance, adjust its parameters as needed, and establish stop-loss orders to mitigate potential losses.

Moreover, AI can play a crucial role in portfolio rebalancing. An AI-powered portfolio management tool can continuously monitor market conditions and adjust the portfolio's asset allocation to maintain the desired risk-return profile. This dynamic rebalancing can help to optimize portfolio performance while minimizing risks. However, the investor should remain involved in the rebalancing process, providing oversight and ensuring that the AI's recommendations align with their long-term investment goals.

The integration of AI into investment decision-making is not without its challenges. One significant hurdle is the "black box" nature of some advanced

AI algorithms. Understanding how these algorithms arrive at their conclusions is crucial for building trust and ensuring the responsible use of AI. Explainable AI (XAI) techniques are emerging to address this challenge, providing more transparent and interpretable insights into the decision-making process. However, XAI is still an evolving field, and further advancements are needed to make complex AI models fully transparent.

Another challenge is the potential for data manipulation and misuse. AI algorithms are only as good as the data they are trained on. If the data is incomplete, inaccurate, or biased, the AI's outputs will be unreliable. It is essential to use high-quality, validated data to train AI models, ensuring that the data reflects the true market conditions. Furthermore, measures should be in place to prevent the manipulation or misuse of AI tools for malicious purposes, such as insider trading or market manipulation.

In conclusion, AI offers significant potential to enhance investment decision-making, but its effective application requires a careful balance between harnessing its analytical power and maintaining human oversight. It's crucial to view AI not as a replacement for human judgment but as a powerful tool to augment our capabilities. By combining AI insights with traditional investment knowledge, sound risk management practices, and a healthy dose of skepticism, investors can leverage the transformative potential of AI to make more informed and successful investment decisions. The future of investing lies in the synergistic relationship between human ingenuity and artificial intelligence, a collaboration that promises to unlock new levels of efficiency and performance in the financial markets. The key to success is a thoughtful, cautious, and ethical integration of AI into established investment strategies.

The integration of AI into traditional financial markets, as discussed previously, presents a compelling case for enhanced investment strategies. However, the unique characteristics of the cryptocurrency market demand a separate, nuanced examination. Cryptocurrencies, with their inherent volatility and decentralized nature, present both unprecedented opportunities and significant risks for AI-driven trading. While the principles of risk management and ethical considerations remain paramount, the application of AI in this domain requires a more cautious and sophisticated approach.

The volatility of the cryptocurrency market is a double-edged sword. On one hand, this volatility creates frequent and substantial price swings, presenting lucrative opportunities for short-term trading strategies powered by AI. On the other hand, the unpredictable nature of these price movements renders traditional forecasting models far less reliable. AI algorithms, trained on historical data, might struggle to accurately predict future price actions in a market as prone to sudden, dramatic shifts as the cryptocurrency market.

One of the most common AI-driven strategies in cryptocurrency trading involves the use of sophisticated algorithmic trading bots. These bots can analyze vast quantities of data in real-time, including price movements, trading volume, social media sentiment, and news articles, to identify potential trading opportunities. They can execute trades at speeds far exceeding human capabilities, capitalizing on fleeting price fluctuations. However, the effectiveness of these bots is heavily dependent on the quality and accuracy of the data they are trained on, as well as the sophistication of their algorithms. A poorly designed bot, trained on insufficient or biased data, could lead to significant losses.

Several AI techniques are employed in these trading bots. Machine learning algorithms, such as recurrent neural networks (RNNs) and long short-term memory (LSTM) networks, are particularly well-suited to analyzing time-series data like cryptocurrency price charts. These algorithms can identify patterns and trends that might be imperceptible to human traders, potentially providing an edge in predicting future price movements. Natural language processing (NLP) is also used to analyze social media sentiment and news articles, gauging market sentiment and identifying potential catalysts for price changes.

However, relying solely on AI-driven trading bots in the cryptocurrency market carries significant risks. The black box nature of many advanced AI algorithms can make it difficult to understand why a particular trade was executed, hindering effective risk management. Furthermore, the decentralized and often unregulated nature of the cryptocurrency market increases the susceptibility to manipulation and fraud. AI algorithms can be vulnerable to attacks designed to exploit their weaknesses and manipulate market prices.

Another important aspect of AI in cryptocurrency trading is the use of AI-powered portfolio management tools. These tools can analyze an investor's risk tolerance, investment goals, and market conditions to construct and manage a diversified cryptocurrency portfolio. They can automatically rebalance the portfolio in response to market changes, helping to optimize returns while minimizing risk. However, the selection of an appropriate portfolio management tool is crucial. It's essential to ensure that the tool employs robust risk management strategies and is transparent in its decision-making processes.

The application of AI in predicting cryptocurrency price movements is an area of ongoing research and development. While some AI algorithms have shown promising results in certain situations, their accuracy remains highly variable and cannot be guaranteed. The inherent unpredictability of the cryptocurrency market makes it challenging to develop AI models that consistently outperform simpler trading strategies. Furthermore, the regulatory environment surrounding cryptocurrencies is constantly evolving, which can impact the accuracy of AI predictions.

Beyond price prediction, AI can play a crucial role in fraud detection and risk mitigation in the cryptocurrency market. AI algorithms can analyze transaction patterns to identify suspicious activities, such as money laundering or pump-and-dump schemes. This can help to protect investors from fraudulent activities and enhance the security of the cryptocurrency ecosystem. However, the development of effective AI-based fraud detection systems requires continuous adaptation to evolving fraud techniques.

The ethical considerations surrounding the use of AI in cryptocurrency trading are particularly critical. The potential for AI to be used for malicious purposes, such as market manipulation or insider trading, is significant. It's essential to ensure that AI-driven trading systems are transparent, auditable, and compliant with all applicable regulations. The development and deployment of AI systems in this context should prioritize ethical considerations and responsible innovation.

In conclusion, AI offers a wide range of opportunities for enhancing trading strategies and risk management in the cryptocurrency market. However, it's crucial to approach this area with a degree of caution and skepticism. The inherent volatility of cryptocurrencies and the potential for manipulation require a more robust and carefully considered approach than in traditional financial markets. Effective use of AI in this domain depends on a combination of sophisticated algorithms, high-quality data, robust risk management, and a strong ethical framework. The responsible and ethical application of AI in cryptocurrency trading remains a key challenge, but one with significant potential to improve both market efficiency and investor protection. Continuous monitoring, adaptation to evolving market dynamics, and a healthy dose of human oversight are indispensable for success in this rapidly evolving field. The future of AI in cryptocurrency trading will likely see further advancements in algorithm design, data analysis, and risk management techniques, but the need for vigilance and ethical considerations will remain paramount. The power of AI lies in its ability to enhance human capabilities, not replace them entirely. This synergistic approach, where human expertise and AI tools work in concert, offers the greatest potential for success in the volatile and complex world of cryptocurrency trading. The key is to leverage

AI's analytical power while maintaining a critical and responsible mindset, understanding its limitations and continually adapting to the ever-changing dynamics of this innovative and rapidly evolving market.

The application of artificial intelligence (AI) is rapidly reshaping the landscape of private equity, significantly impacting the crucial process of due diligence. Traditional due diligence, a labor-intensive and time-consuming undertaking, often relies heavily on manual review of extensive financial documents, legal contracts, and market research. This process, while thorough, is prone to human error and suffers from inherent limitations in the speed and scope of analysis. AI, with its ability to process vast datasets and identify subtle patterns, offers a powerful solution to these challenges, enhancing both the efficiency and accuracy of due diligence in private equity.

One of the most significant impacts of AI in this context is the automation of data collection and analysis. AI-powered tools can rapidly sift through mountains of unstructured data, including news articles, social media posts, SEC filings, and company websites, extracting relevant information and identifying key trends far more quickly than any human team. This capability is particularly valuable in the early stages of due diligence, where a broad overview of the target company and its market is crucial. For example, an AI system can analyze news articles and social media sentiment to gauge public perception of a company, its products, and its management team. This sentiment analysis can provide valuable insights into potential risks or opportunities that might otherwise be overlooked in a traditional due diligence process.

Furthermore, AI can enhance the analysis of financial statements and other structured data. Machine learning algorithms can identify anomalies, inconsistencies, and potential red flags in financial reports that might escape human notice. For instance, an AI system can detect unusual patterns in revenue growth, expense ratios, or cash flow, triggering further investigation by human analysts. This capability helps to mitigate the risk of investing in companies with hidden financial problems. The ability of AI to identify these subtle inconsistencies far surpasses the capabilities of human analysts reviewing massive spreadsheets and reports. This automated anomaly detection saves valuable time and resources while significantly enhancing the accuracy of the due diligence process.

Beyond financial data, AI can also be utilized to analyze legal documents and contracts. Natural Language Processing (NLP) techniques enable AI systems to extract key clauses, identify potential legal risks, and compare contractual terms across multiple agreements. This capability is especially useful in complex transactions involving multiple parties and intricate legal arrangements. For instance, an AI system can quickly scan hundreds of pages of contracts to identify clauses relating to intellectual property rights, liabilities, or termination provisions. This automated analysis frees up legal professionals to focus on more complex aspects of the transaction, significantly accelerating the overall due diligence process.

However, the adoption of AI in private equity due diligence is not without its challenges. One significant concern is the issue of data quality. AI algorithms are only as good as the data they are trained on. Inaccurate or incomplete data can lead to flawed conclusions and potentially disastrous investment decisions. Therefore, rigorous data cleaning and validation are crucial before employing AI tools in due diligence. This need for high-quality data necessitates substantial investment in data infrastructure and expertise.

Another challenge is the "black box" nature of some AI algorithms. While these algorithms can provide insightful predictions, it can be difficult to understand the reasoning behind their conclusions. This lack of transparency can make it challenging to justify investment decisions based on AI-driven insights to stakeholders. Therefore, it's vital to select AI tools that offer some level of explainability and allow for human oversight of the decision-making process. The development of explainable AI (XAI) is crucial to address these concerns and build trust in the use of AI in private equity.

Moreover, the ethical considerations surrounding the use of AI in due diligence cannot be ignored. AI systems must be designed and implemented in a way that avoids biases and ensures fairness. Biased algorithms can lead to discriminatory outcomes, perpetuating existing inequalities in the investment landscape. Therefore, careful consideration must be given to the design and training of AI models to ensure they are free from bias and promote equitable investment practices. Regular audits and monitoring of the AI systems are also necessary to ensure continued fairness and prevent unintended consequences.

Furthermore, the integration of AI into existing due diligence workflows requires careful planning and execution. It's essential to consider the interplay between human expertise and AI tools. AI should be viewed as a powerful augmentation of human capabilities, not a replacement for human judgment. Successful implementation requires a combination of technological expertise and deep understanding of the private equity investment process. A collaborative approach, where AI tools support and augment human analysts, is most effective. Training staff to utilize and interpret the output of AI tools is also essential for maximizing the benefits of AI integration.

Finally, the ongoing development of AI technologies continues to present both opportunities and challenges for private equity due diligence. The advancement of more sophisticated algorithms, improved data processing capabilities, and the development of explainable AI will further enhance the accuracy and efficiency of the process. However, it's equally crucial to maintain a critical perspective, carefully monitoring the evolution of AI techniques and addressing any emerging limitations or ethical concerns. The ongoing evolution of AI will require continuous adaptation and learning, ensuring that the benefits of AI are harnessed responsibly and effectively within the private equity landscape. The future of due diligence in private equity lies not in the replacement of human expertise but in its powerful augmentation by the advanced analytical capabilities of AI, creating a more efficient, accurate, and responsible investment process.

The integration of artificial intelligence (AI) is rapidly transforming various sectors, and the real estate investment market is no exception. The traditional methods of real estate valuation, risk assessment, and market prediction, often reliant on human intuition and limited data analysis, are being augmented, and in some cases replaced, by AI-powered tools. This shift promises increased efficiency, accuracy, and potentially higher returns for investors.

One of the most significant applications of AI in real estate is in property valuation. Traditional appraisal methods are time-consuming and subjective, relying heavily on comparable sales data and expert judgment. AI algorithms, particularly those leveraging machine learning techniques, can analyze vast datasets encompassing not only comparable sales but also numerous other factors influencing property value. These factors might include location data (proximity to schools, parks, transportation hubs, crime rates), property features (size, number of bedrooms and bathrooms, age, condition, amenities), macroeconomic indicators (interest rates, inflation, employment rates), and even local government regulations and zoning laws. By considering this multifaceted data, AI models can generate more accurate and comprehensive valuations than traditional methods, minimizing human bias and reducing appraisal time significantly. For instance, an AI model might identify a subtle trend of increased property values in a specific neighborhood due to upcoming infrastructure improvements, a factor that a human appraiser might overlook or misjudge.

Beyond valuation, AI plays a crucial role in risk assessment. Real estate investments, by their nature, carry inherent risks, including market volatility, changes in local regulations, and unforeseen maintenance costs. AI algorithms can analyze historical data, market trends, and external factors to identify potential risks more effectively than traditional methods. For instance, an AI system can predict the likelihood of a property experiencing flooding based on historical rainfall patterns and proximity to waterways, enabling investors to make more informed decisions about whether to proceed with a purchase. Similarly, it can analyze lease agreements and tenant history to assess the risk of tenant default or vacancy, informing investment strategies and rental pricing.

This predictive capability empowers investors to mitigate potential losses and make more robust investment decisions.

Market prediction is another area where AI is making a significant impact. By analyzing vast amounts of historical data, current market trends, and external factors, AI algorithms can forecast future property price movements and market conditions with a degree of accuracy previously unattainable. This predictive power enables investors to identify opportune moments to buy or sell properties, optimizing their portfolios and maximizing returns. For example, an AI system might predict a downturn in the market for a specific type of property in a particular area, allowing investors to adjust their investment strategy to avoid potential losses. Conversely, it could identify emerging markets with high growth potential, enabling investors to capitalize on early opportunities.

Several AI-powered platforms are already being used in the real estate industry to deliver these functionalities. These platforms often incorporate a combination of machine learning algorithms, natural language processing (NLP), and computer vision. Machine learning algorithms are crucial for analyzing large datasets and identifying patterns, NLP is used for processing unstructured text data such as news articles and market reports, while computer vision enables the automated analysis of images and videos of properties. For example, some platforms use computer vision to automatically assess the condition of a property from photos, providing an objective assessment of its repair needs and overall value. Other platforms leverage NLP to monitor news and social media sentiment related to specific markets, providing insights into emerging trends and potential risks.

However, the application of AI in real estate investment is not without its challenges. One significant limitation is the availability and quality of data. AI algorithms rely on vast amounts of accurate and reliable data to function effectively. In many cases, data in the real estate sector is fragmented, inconsistent, or incomplete, limiting the accuracy and reliability of AI predictions. Another challenge lies in the interpretability of AI models. Many sophisticated AI algorithms, particularly deep learning models, operate as "black boxes," making it difficult to understand the reasoning behind their predictions. This lack of transparency can make it difficult for investors to trust the output of AI systems and justify their decisions based on AI-driven insights.

The potential for bias is another major concern. AI models are trained on historical data, which may reflect existing biases in the real estate market. If not carefully addressed, this can lead to biased predictions that perpetuate inequalities. For example, an AI model trained on data that historically undervalues properties in certain neighborhoods might continue to produce biased valuations, disadvantaging investors or homeowners in those areas. The development of robust methods to mitigate bias and ensure fairness in AI models is therefore crucial.

Furthermore, the ethical implications of using AI in real estate require careful consideration. Concerns about data privacy, algorithmic transparency, and the potential displacement of human appraisers need to be addressed. Regulations and guidelines are needed to ensure responsible and ethical use of AI in this sector.

Despite these challenges, the potential benefits of AI in real estate investment are substantial. As data quality improves and AI algorithms become more sophisticated and transparent, the accuracy and efficiency of AI-driven tools are expected to increase further. The combination of human expertise and AI-powered insights holds the promise of a more efficient, accurate, and equitable real estate investment landscape. Investors who embrace these technologies will likely gain a significant competitive edge in the years to come. The future of real estate investment is likely to be significantly shaped by the ongoing advancements and responsible implementation of AI. This will require a collaborative effort between technology developers, real estate professionals, and policymakers to address the challenges and harness the full potential of this transformative technology. The effective use of AI will not only enhance the efficiency and profitability of real estate investments, but also contribute to a more transparent and equitable market. The successful integration of AI will not replace human judgment but rather augment it, creating a more powerful and insightful approach to real estate investment. By understanding both the opportunities and the limitations, investors can strategically leverage AI to navigate the complexities of the real estate market and achieve superior results.

The application of artificial intelligence (AI) extends beyond the traditional realms of real estate and into the sophisticated world of hedge fund management, fundamentally altering how these firms operate and compete. Hedge funds, known for their aggressive investment strategies and high-stakes environment, are increasingly turning to AI to gain an edge in a highly competitive market. This adoption is driven by the potential for AI to enhance speed, accuracy, and efficiency across various aspects of the investment process, from identifying trading opportunities to managing risk.

One of the most significant applications of AI in hedge fund management is in quantitative trading (quant trading). Quant trading relies on mathematical models and algorithms to identify and execute trades, minimizing human intervention and maximizing speed. AI algorithms, particularly machine learning models, can analyze vast quantities of market data – far exceeding the capacity of human analysts – to identify patterns, trends, and anomalies that might represent lucrative trading opportunities. These algorithms can process news articles, social media sentiment, economic indicators, and company financial statements, identifying correlations and predictive signals that might escape human observation. For example, an AI algorithm might detect a subtle shift in investor sentiment towards a particular sector, allowing a hedge fund to adjust its portfolio before a wider market reaction occurs. Furthermore, AI-powered systems can execute trades at optimal times and prices, capitalizing on fleeting opportunities that would be impossible for humans to exploit. The speed and efficiency of AI-driven quant trading give hedge funds a distinct competitive advantage in the fast-paced world of financial markets.

Beyond identifying opportunities, AI plays a crucial role in portfolio construction and optimization. Traditional portfolio management often relies on diversification and risk-adjusted return models. AI algorithms can go beyond these basic approaches, utilizing sophisticated techniques like reinforcement learning to create portfolios that are dynamically adjusted based on real-time market conditions and risk profiles. This allows hedge funds to adapt their strategies more quickly to changing market dynamics, mitigating potential losses and maximizing returns. The ability of AI to process a multitude of factors simultaneously allows for more nuanced and responsive portfolio management than traditional methods, taking into account factors such as correlation between assets, volatility clustering, and market regime shifts. This advanced portfolio optimization capabilities can significantly improve Sharpe ratios, a key performance indicator for hedge funds.

Risk management is another critical area where AI is making a considerable impact. Hedge funds operate in a high-risk environment, and effective risk management is essential for survival. AI algorithms can analyze historical market data, economic indicators, and other relevant factors to identify potential risks more effectively than traditional methods. For example, AI can predict the likelihood of market crashes or other major events, allowing hedge funds to adjust their portfolios accordingly. Moreover, AI can help hedge funds monitor and manage their exposure to various risk factors, such as credit risk, market risk, and operational risk. This precise risk assessment is crucial to avoiding catastrophic losses. This capability is particularly important in light of the increasing interconnectedness of global financial markets. AI's ability to analyze complex interactions between different asset classes and risk factors enables more robust risk mitigation strategies.

Despite the significant potential of AI in hedge fund management, implementation presents a unique set of challenges. The "black box" nature of some AI algorithms, particularly deep learning models, can be a significant hurdle. The lack of transparency in how these algorithms arrive at their conclusions makes it difficult for fund managers to understand the rationale behind investment decisions. This lack of interpretability can be a barrier to trust and adoption, especially in a field that demands accountability and justification for every move. This opacity can also make it challenging to identify and correct biases within the algorithms themselves. Addressing this challenge often involves employing techniques like explainable AI (XAI), which seeks to make the decision-making process of AI more transparent and understandable.

Data quality and availability pose another major challenge. AI algorithms rely on vast amounts of high-quality data to function effectively. However, financial data can be incomplete, inconsistent, or even inaccurate. This can lead to flawed AI models that produce unreliable predictions. Moreover, the sheer volume of data needed to train sophisticated AI models requires substantial computational resources and expertise in data cleaning and preprocessing. The cost and complexity of acquiring and managing this data represent a significant barrier for some hedge funds. The quality of data is crucial for the accuracy of AI models; noisy or biased data can lead to poor predictions and costly investment decisions.

The competitive landscape also plays a role in the adoption of AI in hedge fund management. The industry is extremely competitive, and even a slight edge can make a significant difference in performance. The firms that are able to effectively leverage AI are likely to gain a significant competitive advantage. However, the adoption of AI is not uniform; some hedge funds are leading the way, while others are lagging behind. The pace of adoption will likely depend on factors such as the size of the fund, access to technology and talent, and risk tolerance. The potential for AI to disrupt the industry is significant, and the firms that are able to adapt and innovate are likely to thrive.

Another critical challenge lies in the integration of AI into existing workflows and infrastructure. Hedge funds often have legacy systems and established processes, and integrating AI can be a complex and time-consuming process. It requires significant investment in technology, infrastructure, and skilled personnel. This can be a barrier to entry for smaller funds that lack the resources to invest in AI-powered systems. Furthermore, the organizational culture within a hedge fund will influence its ability to successfully adopt and integrate AI. A culture of innovation and a willingness to embrace new technologies are essential.

Ethical considerations are also paramount. AI algorithms can perpetuate biases present in the data they are trained on, potentially leading to unfair or discriminatory outcomes. For instance, an algorithm trained on historical data that reflects gender or racial biases could make biased investment decisions. Hedge funds need to be mindful of these ethical implications and take steps to mitigate bias and ensure fairness in their AI systems. This requires careful consideration of data selection, algorithm design, and ongoing monitoring of AI systems for bias. Regulations and guidelines are also needed to ensure responsible and ethical use of AI in the finance sector. Transparency and accountability are key aspects of ethical AI implementation.

Despite these challenges, the future of hedge fund management is inextricably linked to AI. The benefits of enhanced speed, accuracy, and efficiency are too significant to ignore. As AI technologies continue to evolve and become more accessible, their adoption in the hedge fund industry is likely to accelerate. The firms that are able to successfully integrate AI into their operations will gain a significant competitive advantage, while those that fail to adapt risk being left behind. The continued development of explainable AI (XAI) will help address concerns about the "black box" nature of certain AI algorithms, making AI-driven investment decisions more transparent and understandable. Ultimately, the successful application of AI will involve a combination of advanced technology, skilled personnel, and a proactive approach to addressing the inherent challenges. The future of hedge fund management is not about replacing human expertise but rather augmenting it with the power of AI.

The integration of AI into alternative investment strategies presents a compelling case for enhanced efficiency and potentially higher returns. However, this technological leap necessitates a rigorous examination of the ethical implications inherent in its application. The very power of AI to analyze vast datasets and identify complex patterns also carries the risk of amplifying existing biases or creating new ones, leading to unfair or discriminatory outcomes. This section delves into these crucial ethical considerations, providing a framework for responsible AI development and deployment within the alternative investment landscape.

One of the most significant ethical concerns revolves around algorithmic bias. AI models are trained on historical data, and if this data reflects societal biases—be it racial, gender, or socioeconomic—the resulting algorithm will likely perpetuate these biases in its predictions and investment decisions. For instance, an AI system trained on lending data that historically favored certain demographic groups might inadvertently discriminate against others, perpetuating existing inequalities. In the context of alternative investments, this could manifest in various ways. An AI-driven private equity fund, for example, might inadvertently overlook promising ventures from underrepresented founders due to biases embedded in its training data. Similarly, an AI system assessing real estate investment opportunities might undervalue properties in certain neighborhoods based on historical data reflecting discriminatory lending practices.

The problem is exacerbated by the "black box" nature of some AI algorithms, particularly deep learning models. The lack of transparency in how these algorithms arrive at their conclusions makes it difficult to identify and address biases. While the algorithms themselves may not be inherently biased, the data they are trained on almost certainly contains biases reflecting historical realities. This lack of transparency makes it challenging to hold AI systems accountable for their decisions and to ensure fairness and equity in the investment process. Understanding the "why" behind an AI's recommendation is critical, not only for regulatory compliance but also for investor confidence and trust. The absence of clear explanations can erode confidence in the system's objectivity and fairness.

Addressing this challenge necessitates the adoption of explainable AI (XAI) techniques. XAI aims to make the decision-making process of AI more transparent and understandable, allowing investors and regulators to scrutinize the reasoning behind investment recommendations. This involves developing methods to interpret and explain the complex internal workings of AI algorithms, providing insights into the factors influencing their decisions. While XAI is still an evolving field, its implementation is crucial for building trust and ensuring responsible use of AI in alternative investments. This is not merely an ethical imperative; it's also a business imperative. Investors are increasingly demanding transparency and accountability in their investments, particularly those involving complex AI-driven strategies.

Beyond algorithmic bias, the issue of data privacy and security is paramount. AI algorithms require access to vast amounts of data, often including sensitive personal and financial information. The responsible handling of this data is crucial to protecting the privacy and security of individuals and organizations. Alternative investment strategies often involve detailed financial information, proprietary investment strategies, and confidential business intelligence. This data needs to be protected from unauthorized access, use, or disclosure. Robust data security measures are essential, including encryption, access controls, and regular security audits. Compliance with relevant data privacy regulations, such as GDPR and CCPA, is also non-negotiable.

Furthermore, the potential for manipulation and misuse of AI in alternative investments should not be underestimated. Malicious actors could attempt to manipulate data sets used to train AI algorithms, leading to inaccurate or biased predictions that benefit them at the expense of others. Robust data validation and verification procedures are essential to mitigate this risk. Regular audits of the data and algorithms used in AI-driven investment strategies are crucial to ensuring their integrity and prevent malicious manipulation.

The regulatory landscape surrounding AI in alternative investments is still developing. However, clear guidelines and regulations are essential to ensure the responsible and ethical use of AI in this space. Regulators need to develop frameworks that address algorithmic bias, data privacy, and security concerns. These regulations should balance the need for innovation with the protection of investors and the broader financial system. The lack of comprehensive regulatory oversight can create a breeding ground for unethical practices, hindering the wider adoption of AI in the industry. Collaboration between regulators, industry players, and AI researchers is needed to develop effective and balanced regulatory frameworks.

The ethical considerations also extend to the potential displacement of human workers. As AI systems become more sophisticated, they could automate tasks previously performed by human analysts and portfolio managers. This raises concerns about job displacement and the need for workforce retraining and adaptation. Responsible AI deployment necessitates a thoughtful approach to managing the human element. This involves investing in education and training programs to equip workers with the skills needed to navigate the changing job market. It also means considering the social and economic implications of AI-driven automation and developing strategies to mitigate potential negative impacts on employment.

In conclusion, the application of AI in alternative investments offers tremendous opportunities for improved efficiency, higher returns, and innovative investment strategies. However, the ethical implications cannot be overlooked. Addressing algorithmic bias, ensuring data privacy and security, promoting transparency through XAI, and developing robust regulatory frameworks are crucial for fostering trust and responsible innovation in this rapidly evolving field. A proactive approach to ethical considerations is not just a matter of compliance; it's a key to unlocking the full potential of AI while safeguarding the integrity and fairness of the alternative investment landscape. The future of alternative investing will be defined not only by technological advancements but also by the ethical frameworks that guide their deployment. A future where AI enhances, rather than replaces, human expertise and judgment will require a sustained commitment to ethical AI principles and ongoing dialogue between stakeholders.

The democratization of finance, a long-held aspiration, is rapidly becoming a reality thanks to the transformative power of artificial intelligence. While previously accessible primarily to high-net-worth individuals, sophisticated investment strategies are now within reach of a broader population through AI-driven personalized investment advice. This personalized approach leverages AI's ability to analyze vast datasets – encompassing market trends, individual risk profiles, financial goals, and even behavioral patterns – to craft customized portfolios tailored to each client's unique circumstances. This represents a significant shift from the traditional, often one-size-fits-all approach to investment management.

The process begins with a comprehensive assessment of the client's financial situation, including income, assets, liabilities, and investment goals. AI algorithms then analyze this data in conjunction with market intelligence and historical performance data to identify optimal investment strategies. Unlike human advisors who might be limited in the number of clients they can serve effectively, AI can simultaneously manage thousands of portfolios, providing consistent, high-quality advice at scale. This efficiency translates to reduced costs for clients, making sophisticated investment management more affordable and accessible.

Furthermore, AI-powered platforms can continuously monitor market conditions and adjust investment portfolios dynamically. This real-time adaptation allows for quick responses to market fluctuations, potentially mitigating losses and maximizing returns. Traditional investment strategies often involve periodic rebalancing, which may lag behind rapidly shifting market dynamics. AI, on the other hand, can react instantaneously to changing conditions, providing a level of responsiveness that is unattainable with human-only management.

Beyond portfolio construction and management, AI is also revolutionizing client engagement and financial education. Chatbots and virtual assistants powered by natural language processing (NLP) can answer client queries, provide explanations of complex financial concepts, and offer personalized financial guidance 24/7. This accessibility fosters greater financial literacy and empowers clients to make informed decisions about their investments. AI can also personalize communication, tailoring its messaging to resonate with individual client preferences and understanding levels.

However, the integration of AI into personalized investment advice also presents significant challenges and ethical considerations. One primary concern revolves around data privacy and security. AI algorithms require access to sensitive personal and financial information, raising concerns about potential breaches and misuse of data. Robust security measures, including encryption, access controls, and regular security audits, are crucial to mitigating these risks. Compliance with data privacy regulations, such as GDPR and CCPA, is non-negotiable, demanding a rigorous approach to data governance and transparency.

Another significant challenge lies in algorithmic bias. AI models are trained on historical data, which may reflect existing societal biases. If these biases are not adequately addressed, the AI system could perpetuate inequalities by disproportionately favoring certain client demographics. For example, an AI system trained on lending data that historically discriminated against specific groups might inadvertently make biased investment recommendations, perpetuating historical inequities. Mitigating algorithmic bias requires careful curation of training data, rigorous testing for bias, and the implementation of explainable AI (XAI) techniques to make the decision-making process transparent and accountable. Ongoing monitoring and auditing of AI systems are essential to identify and address emerging biases.

The "black box" nature of some AI algorithms further complicates matters. Understanding why an AI system makes a specific recommendation is crucial for both clients and regulators. The lack of transparency in complex AI models can erode client trust and hinder regulatory oversight. XAI techniques aim to address this challenge by providing insights into the reasoning behind AI-driven investment decisions. However, XAI is still an evolving field, and its effective implementation requires significant research and development.

Furthermore, the potential for manipulation and misuse of AI in personalized investment advice must not be underestimated. Malicious actors could attempt to manipulate data or algorithms to gain an unfair advantage. Robust data validation and verification procedures, coupled with regular audits of AI systems, are essential to safeguarding against such threats. Collaboration between AI developers, financial institutions, and regulators is vital to establishing robust security and integrity standards.

The regulatory landscape surrounding AI-driven investment advice is still evolving, creating a need for clear guidelines and regulations. These regulations should address concerns about data privacy, algorithmic bias, and transparency. A balanced approach is necessary, promoting innovation while ensuring the protection of investors. Regulatory bodies need to adapt to the rapid advancements in AI, establishing frameworks that maintain investor confidence and prevent the misuse of technology. International collaboration is critical to ensure consistent standards across different jurisdictions.

Beyond technical and regulatory challenges, the ethical implications of AI in personalized investment advice extend to the potential displacement of human financial advisors. As AI systems become more sophisticated, they may automate tasks previously performed by human professionals. This necessitates a thoughtful approach to workforce transition, investing in retraining and upskilling programs to equip financial professionals with the skills needed to thrive in a changing landscape. The responsible integration of AI should augment human expertise, not replace it entirely, focusing on leveraging human judgment and empathy in areas where AI falls short.

In conclusion, AI is poised to revolutionize the landscape of personalized investment advice, making sophisticated investment strategies accessible to a broader audience. However, realizing the full potential of AI requires careful consideration of ethical implications, including data privacy, algorithmic bias, transparency, and the impact on the workforce. Robust security measures, transparent algorithms, and responsible regulation are crucial to fostering trust and ensuring the ethical application of AI in the financial services sector. A future where AI enhances, rather than replaces, human expertise requires ongoing dialogue between developers, regulators, and the broader financial community. This collaborative approach will be key to unlocking the transformative power of AI while safeguarding the interests of investors and maintaining the integrity of the financial system.

The transformative potential of AI extends far beyond traditional centralized finance, reaching into the burgeoning world of decentralized finance (DeFi). DeFi, built upon blockchain technology, promises to democratize financial services by eliminating intermediaries and empowering individuals to control their own assets. However, the complexity and volatility inherent in DeFi ecosystems present significant challenges for users. This is where AI steps in, offering the potential to enhance the efficiency, security, and accessibility of DeFi platforms.

AI can significantly improve the user experience within DeFi. Navigating the multitude of decentralized applications (dApps), understanding complex smart contracts, and managing risk effectively can be daunting for even experienced users. AI-powered tools can simplify these processes, providing intuitive interfaces, personalized recommendations, and automated risk management. Imagine a DeFi dashboard powered by AI that automatically aggregates portfolio balances across various dApps, identifies potential risks based on real-time market data, and suggests optimized investment strategies based on the user's risk tolerance and financial goals. This level of automation and personalized support could significantly lower the barrier to entry for newcomers to the DeFi space, fostering broader adoption and participation.

One of the core functions of DeFi is lending and borrowing. AI can revolutionize this process through improved credit scoring and risk assessment. Traditional credit scoring systems often rely on limited data points and may exclude individuals who lack a formal credit history. In contrast, AI can analyze a much broader range of on-chain data, such as transaction history, smart contract interactions, and even social media activity, to create more comprehensive and inclusive credit scoring models. This could unlock access to credit for individuals who have been historically underserved by traditional financial institutions. Moreover, AI can optimize lending rates by dynamically adjusting them based on real-time risk assessments and market conditions, leading to more efficient allocation of capital within the DeFi ecosystem.

AI's role in algorithmic trading within DeFi is also transformative. DeFi protocols offer a wide range of trading opportunities, but navigating the complexities of decentralized exchanges (DEXs) and identifying profitable trading strategies requires significant expertise. AI-powered trading bots can analyze market data, identify arbitrage opportunities, execute trades autonomously, and optimize portfolio performance. These bots can operate 24/7, reacting instantly to market fluctuations and exploiting fleeting opportunities that might be missed by human traders. However, the deployment of AI trading bots also raises concerns about market manipulation and the potential for exacerbating volatility. Robust safeguards and regulatory oversight are essential to ensure the ethical and responsible use of AI in DeFi trading.

Beyond trading, AI can enhance the security of DeFi platforms. Smart contracts, the backbone of DeFi, are susceptible to vulnerabilities that malicious actors can exploit. AI can assist in identifying and mitigating these vulnerabilities through automated security audits and penetration testing. AI models can analyze smart contract code for potential bugs and vulnerabilities, flagging potential risks before they can be exploited. Moreover, AI can assist in detecting and preventing fraudulent activities within DeFi, such as wash trading or rug pulls. Real-time monitoring of on-chain transactions and identification of anomalous patterns can provide early warnings of suspicious activity. This proactive approach can help to protect users and maintain the integrity of the DeFi ecosystem.

However, the integration of AI into DeFi is not without its challenges. Data privacy and security remain paramount concerns. AI algorithms require access to significant amounts of on-chain data, raising concerns about potential misuse of sensitive information. Robust encryption, secure data storage, and compliance with data privacy regulations are essential to mitigating these risks. Furthermore, the "black box" nature of some AI algorithms can make it difficult to understand how decisions are made, raising transparency and accountability concerns. Explainable AI (XAI) techniques can help address this challenge, but the development and implementation of XAI remain ongoing areas of research and development.

Algorithmic bias is another crucial concern. AI models are trained on historical data, which may reflect existing biases within the DeFi ecosystem. If not properly addressed, these biases can perpetuate inequalities and lead to unfair outcomes. For example, an AI-powered lending platform trained on biased data might unfairly deny credit to certain groups of users. Mitigating algorithmic bias requires careful data curation, rigorous testing for bias, and the implementation of fairness-aware algorithms. Continuous monitoring and auditing of AI systems are crucial to identify and address emerging biases.

The regulatory landscape for AI in DeFi is still evolving, creating a need for clear guidelines and regulations. These regulations should address concerns about data privacy, algorithmic bias, security vulnerabilities, and market manipulation. A balanced approach is necessary, promoting innovation while ensuring the protection of users. Regulatory bodies must adapt to the rapid advancements in AI and DeFi, establishing frameworks that maintain user confidence and prevent the misuse of technology. International collaboration is critical to ensuring consistent standards across different jurisdictions.

The future of AI in DeFi holds immense promise. AI-powered tools have the potential to significantly enhance the efficiency, security, and accessibility of DeFi platforms, fostering broader adoption and empowering individuals to participate more fully in the decentralized financial ecosystem. However, realizing this potential requires careful consideration of the ethical implications, including data privacy, algorithmic bias, transparency, and the need for robust regulatory frameworks. A collaborative effort between AI developers, DeFi protocols, regulators, and the broader community is crucial to navigate the challenges and unlock the transformative power of AI in DeFi, ensuring a secure, fair, and inclusive decentralized financial future. This collaboration will involve ongoing research into XAI, development of more robust security measures, and a continuous refinement of regulatory guidelines to keep pace with the rapidly evolving landscape. The goal is not simply to integrate AI into DeFi, but to integrate it responsibly, ethically, and sustainably. Only through such a careful and considered approach can the full benefits of AI be harnessed to unlock the transformative potential of DeFi for everyone. This includes fostering financial inclusion by creating more accessible and transparent systems that cater to a broader range of users, irrespective of their technological expertise or previous financial experience. The democratization of finance, through the combined power of DeFi and AI, promises a future where financial services are more accessible, efficient, and equitable for all.

The integration of AI into the financial sector is rapidly transforming how financial institutions operate and interact with their clients. However, this rapid evolution presents significant challenges for regulatory bodies tasked with ensuring the safety, security, and fairness of financial markets. The development of robust regulatory frameworks capable of keeping pace with these technological advancements is critical. The inherent complexity of AI algorithms, coupled with their potential for both beneficial and detrimental outcomes, necessitates a nuanced and adaptive regulatory approach.

One primary concern revolves around data privacy. AI algorithms rely heavily on large datasets, often encompassing sensitive personal and financial information. Protecting this data from unauthorized access, use, and disclosure is paramount. Regulations like the General Data Protection Regulation (GDPR) in Europe and the California Consumer Privacy Act (CCPA) in the United States aim to establish standards for data protection, but their application to the novel context of AI in finance requires careful consideration and often necessitates bespoke interpretations. The challenge lies not only in ensuring compliance but also in maintaining the effectiveness of AI systems while safeguarding individual privacy rights. Techniques such as differential privacy and federated learning, which allow for analysis of data without direct access to individual records, represent promising solutions but need wider implementation and standardization.

Algorithmic bias presents another significant hurdle. AI models are trained on historical data, which may reflect existing societal biases. If left unchecked, these biases can be amplified and perpetuated by AI systems, leading to discriminatory outcomes in areas like loan applications, credit scoring, and investment recommendations. Regulatory bodies are increasingly focusing on developing methods for detecting and mitigating algorithmic bias, encompassing both pre- and post-deployment monitoring. This involves not only assessing the fairness of algorithms but also understanding their potential impact on various demographic groups. Transparency requirements, which force greater explainability of AI decision-making processes, are crucial in identifying and addressing potential biases. The development and adoption of fairness-aware algorithms, specifically designed to mitigate bias, are equally important.

The "black box" nature of many AI algorithms poses a further challenge for regulators. The opacity of these systems often makes it difficult to understand how decisions are made, hindering effective oversight and accountability. This lack of transparency makes it challenging to determine whether an AI system is functioning correctly, complying with regulations, or exhibiting unintended biases. The emergence of explainable AI (XAI) offers a potential solution, aiming to make AI decision-making processes more transparent and understandable. However, XAI is still an evolving field, and the development of standardized methods for evaluating the explainability of AI systems remains an ongoing challenge. Regulators need to work closely with AI developers to create standardized reporting requirements and metrics that allow for effective assessment of AI transparency.

Market manipulation presents another significant risk. The use of AI-powered trading bots can potentially amplify market volatility or even facilitate fraudulent activities. The speed and scale at which these bots can operate require new regulatory approaches to detect and prevent manipulation. This involves monitoring market activity for unusual patterns, collaborating with exchanges to implement robust surveillance systems, and developing methods to identify and penalize manipulative behavior. The inherent complexity of AI-driven trading strategies often makes detection challenging, necessitating continuous adaptation of regulatory strategies.

The speed of technological advancement in AI presents a significant challenge for regulatory bodies. Regulations need to be flexible and adaptable to accommodate rapid innovation while maintaining their effectiveness. This often requires a collaborative approach between regulators and AI developers, fostering an environment of responsible innovation and minimizing unintended consequences. "Regulatory sandboxes," which provide controlled environments for testing and evaluating new technologies, represent a promising approach to enabling innovation while ensuring adequate oversight. These sandboxes allow developers to experiment with new AI applications within a structured regulatory framework, enabling regulatory bodies to learn from real-world applications and refine their approach as needed.

The international nature of financial markets further complicates the regulatory landscape. AI applications often transcend national borders, necessitating international cooperation and harmonization of regulatory standards. This requires effective communication and collaboration between different regulatory bodies, preventing regulatory arbitrage and ensuring consistent levels of protection across jurisdictions. International agreements and frameworks can foster a more coordinated and effective regulatory approach to AI in finance.

The regulatory landscape surrounding AI in finance is dynamic and evolving. While existing regulations provide a foundation, they need to be continually adapted to address the unique challenges posed by AI. A collaborative approach, involving regulators, AI developers, financial institutions, and other stakeholders, is crucial to developing a regulatory framework that promotes innovation while safeguarding the stability and integrity of the financial system. This collaborative effort should emphasize transparency, accountability, and fairness, ensuring that AI is used responsibly and ethically in the financial sector for the benefit of all stakeholders. The focus should be on creating regulations that are both effective and adaptable, allowing for innovation while protecting consumers and maintaining the stability of the financial system. This ongoing dialogue and collaboration will be essential to navigate the complex interplay between technological advancement and regulatory oversight, ensuring a secure, fair, and efficient future for the financial sector. The ultimate goal is to harness the power of AI for positive societal impact, promoting financial inclusion, efficiency, and responsible innovation.

Blockchain technology, with its inherent features of decentralization, immutability, and transparency, offers a compelling solution to several challenges posed by the integration of AI in finance. The synergy between these two transformative technologies promises to create a more secure, efficient, and trustworthy financial ecosystem. One of the most significant contributions of blockchain lies in its ability to enhance the security and transparency of AI algorithms. Traditional AI systems often rely on centralized databases, making them vulnerable to hacking, data breaches, and manipulation. Blockchain, however, provides a distributed ledger, meaning that the data is not stored in a single location but is replicated across a network of computers. This eliminates single points of failure and significantly increases the resilience of the system against cyberattacks. Furthermore, the immutability of blockchain ensures that once data is recorded, it cannot be altered or deleted without detection, thus preserving the integrity of the AI's data and operational history.

The transparency afforded by blockchain is another key advantage. The ability to publicly audit the entire lifecycle of an AI algorithm, from its training data to its operational decisions, fosters greater trust and accountability. This is particularly crucial in addressing concerns surrounding algorithmic bias and the "black box" nature of many AI systems. By recording all stages of an algorithm's development and deployment on a public blockchain, stakeholders can scrutinize its operations and ensure fairness and compliance with regulations. This heightened transparency can significantly improve the public's confidence in AI-driven financial services. For example, in the realm of algorithmic trading, a blockchain-based system could record every trade executed by an AI algorithm, making it easier to detect and prevent market manipulation.

The application of blockchain extends beyond the enhancement of security and transparency; it also provides a robust framework for managing and auditing AI algorithms. Consider the complexity involved in managing multiple AI models deployed across various financial institutions. Blockchain can facilitate efficient management of these models by creating a secure and transparent registry. Each AI model can be uniquely identified on the blockchain, along with its associated metadata, such as its version number, training data, performance metrics, and audit logs. This creates a single source of truth for all AI models, streamlining their management and simplifying the regulatory compliance process. Furthermore, the decentralized nature of blockchain ensures that no single entity has control over the registry, fostering a level playing field for all participants.

Smart contracts, self-executing contracts with the terms of the agreement directly written into code, represent another powerful application of blockchain in AI-powered finance. Smart contracts can automate various financial processes, such as loan origination, insurance claims processing, and investment management. By integrating AI algorithms into smart contracts, we can create highly sophisticated and efficient systems that can adapt to changing market conditions and individual needs. For instance, an AI-powered smart contract could automatically adjust the interest rate on a loan based on the borrower's credit score and market conditions, ensuring a fairer and more dynamic lending process. The immutability and transparency of blockchain ensure that these smart contracts are executed precisely as programmed, minimizing the risk of fraud and errors.

However, the integration of blockchain and AI in finance is not without its challenges. The scalability of blockchain technology is a significant concern, especially when dealing with the massive datasets used by sophisticated AI models. Current blockchain implementations may struggle to handle the volume of transactions required for large-scale AI applications. Research and development efforts are focusing on improving the scalability of blockchain through various techniques such as sharding and layer-two solutions. Another challenge lies in the interoperability of different blockchain platforms. Many different blockchain networks exist, each with its own protocols and standards. Lack of interoperability can hinder the seamless integration of AI systems across different platforms. Efforts are underway to establish interoperability standards to facilitate communication and data exchange between different blockchain networks.

Furthermore, the energy consumption of certain blockchain networks remains a concern. Some blockchain technologies, particularly those based on proof-of-work consensus mechanisms, require significant computational power, leading to high energy consumption. This environmental impact needs to be addressed through the adoption of more energy-efficient consensus mechanisms, such as proof-of-stake. The regulatory landscape surrounding the use of blockchain and AI in finance is still evolving. Regulatory clarity is necessary to foster innovation while ensuring the safety and security of the financial system. Collaborations between regulators, AI developers, and blockchain experts are crucial to establishing effective and adaptable regulatory frameworks.

Despite these challenges, the potential benefits of integrating blockchain and AI in finance are undeniable. The combination of blockchain's security and transparency with AI's analytical power can revolutionize various aspects of the financial industry, from risk management and fraud detection to investment strategies and regulatory compliance. The development of decentralized autonomous organizations (DAOs), which are self-governing entities operating on blockchain, presents another exciting opportunity. DAOs can use AI to manage their operations and decision-making processes, fostering greater transparency and accountability. The increased efficiency and reduced costs associated with using blockchain and AI can lead to greater financial inclusion, particularly for underserved communities that lack access to traditional financial services.

For example, microfinance institutions could leverage AI-powered systems on blockchain to assess creditworthiness more effectively and provide loans to individuals who might otherwise be excluded from traditional lending programs. Similarly, cross-border payments could be significantly streamlined through the use of blockchain and AI, reducing costs and processing times. The use of AI-powered predictive models on blockchain can enhance fraud detection and prevention by analyzing vast amounts of transaction data to identify suspicious patterns. This can help financial institutions to protect themselves from financial crimes and safeguard their customers' funds. In the investment world, AI-powered portfolio management tools could be deployed on blockchain to optimize investment strategies based on market conditions and individual risk profiles. This could lead to more efficient and profitable investment outcomes.

The synergy between blockchain and AI represents a transformative paradigm shift in finance. By addressing the challenges and fostering collaborations between stakeholders, we can unlock the full potential of these technologies to create a more secure, efficient, transparent, and inclusive financial system. The future of finance is undoubtedly intertwined with the convergence of blockchain and AI, promising a new era of innovation and opportunity. Continued research and development, coupled with thoughtful regulatory oversight, are crucial to navigating the challenges and realizing the significant potential of this transformative combination. The benefits of enhanced security, increased transparency, improved efficiency, and greater financial inclusion are too significant to ignore. The ongoing evolution and refinement of these technologies, along with the development of appropriate regulatory frameworks, will shape the future of the financial industry for years to come. This collaboration will be key to ensuring responsible innovation and the creation of a fair and equitable financial system for all stakeholders.

The convergence of artificial intelligence (AI) and financial markets is rapidly accelerating, ushering in an era of unprecedented sophistication and efficiency. Beyond the foundational applications already transforming the industry, several emerging trends are poised to reshape the landscape in profound ways. These advancements are not isolated developments but rather interconnected elements contributing to a holistic evolution of the financial ecosystem.

One area of significant progress is the application of Natural Language Processing (NLP) to financial news analysis. Traditional methods of market analysis relied heavily on human interpretation of vast quantities of news articles, reports, and social media commentary. This approach is inherently slow, prone to bias, and limited in scale. NLP, however, enables computers to process and understand human language, extracting key insights from unstructured data at an unparalleled speed and accuracy. Sophisticated NLP algorithms can now identify sentiment shifts, predict market trends based on news narratives, and even identify potentially misleading or manipulative information. This capability is particularly valuable in high-frequency trading (HFT), where milliseconds can make a significant difference in profitability. Algorithms can scan thousands of news sources concurrently, instantly reacting to significant events and adjusting trading strategies accordingly. The enhanced speed and accuracy of NLP-driven analysis translates to more informed decisions, reduced risk, and improved investment returns.

Moreover, the application of NLP is not confined to merely analyzing textual data; it's extending into the realm of audio and video analysis. Financial news broadcasts, earnings calls, and even informal discussions among market participants can now be automatically transcribed and analyzed for sentiment and predictive signals. This multi-modal approach provides a richer and more comprehensive understanding of market dynamics, potentially unveiling subtle clues that might otherwise be missed. The combination of NLP with other AI techniques, like machine learning (ML), is creating even more powerful analytical tools. For instance, ML algorithms can be trained on vast datasets of news articles and corresponding market movements to predict future price fluctuations with increasing accuracy.

Another crucial area experiencing rapid advancement is fraud detection. The increasing complexity of financial transactions and the rise of sophisticated cyberattacks necessitate more robust and adaptive security measures. Traditional fraud detection systems often rely on rule-based approaches, which are inflexible and easily circumvented by cunning fraudsters. Advanced machine learning techniques, particularly deep learning, offer a more powerful solution. Deep learning algorithms can analyze massive datasets of transactions, identifying subtle patterns and anomalies that indicate fraudulent activity. These algorithms continuously learn and adapt, improving their accuracy over time and staying ahead of evolving fraud techniques. This proactive approach significantly enhances the ability of financial institutions to detect and prevent fraud, protecting both the institutions themselves and their customers.

The effectiveness of deep learning in fraud detection is significantly amplified by the integration of other data sources. For example, incorporating geolocation data, social media activity, and even behavioral biometrics can create a more holistic view of each transaction. This multi-faceted approach allows for a more comprehensive risk assessment, further reducing the likelihood of fraudulent transactions slipping through the cracks. Furthermore, the explainability of deep learning models is becoming increasingly important, as regulators and stakeholders demand transparency in AI-driven decision-making. Researchers are actively developing techniques to make these complex models more interpretable, providing insights into the rationale behind their fraud detection decisions.

Beyond NLP and enhanced fraud detection, the integration of AI with quantum computing holds immense potential for the financial industry. Quantum computing leverages the principles of quantum mechanics to perform calculations far beyond the capabilities of classical computers. This technology is still in its early stages of development, but its potential applications in finance are enormous. One area of particular interest is portfolio optimization. Classical algorithms struggle to optimize large and complex portfolios effectively, particularly when considering numerous factors like risk tolerance, diversification goals, and market dynamics. Quantum algorithms, however, can tackle these problems far more efficiently, potentially leading to significantly improved investment strategies.

Furthermore, quantum computing could revolutionize risk management. The ability to simulate complex financial models with unprecedented accuracy could drastically reduce uncertainty and improve risk assessment. This would be invaluable in areas like derivatives pricing, credit risk modeling, and market volatility prediction. The integration of AI and quantum computing is still in its nascent stages, but early research suggests the possibility of developing new algorithms capable of solving previously intractable problems in finance. This synergy could fundamentally alter the way financial institutions manage risk and make investment decisions.

Looking ahead, several future directions in AI for financial markets stand out. One critical area is the development of more robust and explainable AI models. As AI systems become increasingly complex and influential in financial decision-making, the need for transparency and interpretability becomes paramount. This requires the development of new algorithms and techniques that not only provide accurate predictions but also offer clear insights into their reasoning processes. This will build trust and confidence in AI-driven systems, allowing for wider adoption and greater acceptance.

Another key area is the development of AI systems that can adapt to changing market conditions and evolving regulatory frameworks. Financial markets are dynamic environments, constantly subject to unpredictable shifts and new regulations. AI systems need to be capable of adjusting their strategies and models to respond effectively to these changes. This requires the development of more adaptive and robust AI algorithms, potentially leveraging reinforcement learning and other advanced techniques.

The ethical considerations surrounding the use of AI in finance are also of paramount importance. Concerns about algorithmic bias, fairness, and accountability need to be carefully addressed to ensure the responsible and equitable use of AI technologies. Developing mechanisms for detecting and mitigating bias in AI algorithms is a crucial step in building trust and preventing unintended consequences. Furthermore, establishing clear guidelines and regulations for the development and deployment of AI systems in finance is essential to protect consumers and maintain the stability of the financial system.

The integration of AI in financial markets is not merely a technological advancement; it's a transformative shift that is reshaping the industry in fundamental ways. The trends discussed above are just the beginning. As AI technologies continue to evolve, their impact on financial markets will only intensify. The future of finance is inextricably linked to the ongoing development and responsible implementation of AI, promising both unprecedented opportunities and significant challenges that require careful navigation. The ability to harness the power of AI while mitigating its potential risks will be crucial for the success and stability of the financial system in the years to come. This journey requires a collaborative effort between technologists, regulators, and financial professionals to ensure that AI is used ethically, effectively, and responsibly to create a more robust, efficient, and equitable financial ecosystem.

This case study examines a specific instance of successful AI-driven portfolio management, highlighting the practical application of the theoretical frameworks explored in previous chapters. The portfolio in question, managed by a boutique investment firm specializing in quantitative strategies, demonstrated consistent outperformance compared to relevant benchmarks over a five-year period. Its success can be attributed to a multifaceted approach that integrated cutting-edge AI techniques with traditional investment principles.

The core of the strategy involved a proprietary AI model built upon a deep learning architecture. This model was trained on an extensive dataset comprising historical market data, macroeconomic indicators, company financials, news sentiment analysis derived from NLP techniques, and alternative data sources such as satellite imagery and social media trends. The sheer volume of data processed was staggering, far exceeding the capacity of human analysts to handle effectively. This data richness allowed the model to identify subtle correlations and patterns invisible to human observers, providing a significant edge in predicting market movements. The model wasn't simply predicting price movements; it was assessing the probability of various market scenarios, incorporating factors such as geopolitical events, regulatory changes, and technological disruptions.

The portfolio's construction leveraged a factor-based approach, but with a crucial distinction. Traditional factor-based models typically rely on pre-defined factors like value, momentum, and size. In this instance, the AI model dynamically identified and weighted these factors, adjusting its allocations in response to evolving market conditions. For example, during periods of heightened market volatility, the model might increase its exposure to low-volatility stocks and reduce its reliance on momentum strategies. Conversely, during periods of strong economic growth, it might shift its emphasis towards growth stocks and cyclical sectors. This adaptive capability was a key differentiator, allowing the portfolio to navigate market fluctuations more effectively than static factor models.

Furthermore, the AI model didn't simply focus on asset allocation; it actively managed risk. Traditional risk management techniques often rely on historical volatility metrics, which can be misleading during periods of regime change. The AI model, however, incorporated more sophisticated risk measures, drawing upon both historical data and real-time information to assess potential downside risks. It used advanced techniques like stress testing and scenario analysis, simulating various market events and evaluating the portfolio's resilience under different stress conditions. This proactive risk management approach significantly reduced the portfolio's overall risk profile while maintaining its impressive returns.

The AI model's outputs were not merely interpreted by human portfolio managers; they were integrated directly into the trading system. This automated execution process ensured rapid and efficient response to changing market conditions. The system was designed to execute trades based on the model's recommendations, minimizing human intervention and preventing emotional biases from influencing investment decisions. This automation didn't eliminate human oversight altogether; it merely shifted the focus from tactical decision-making to strategic oversight and risk management. Human analysts were responsible for monitoring the AI model's performance, validating its inputs, and ensuring its alignment with the overall investment strategy.

The portfolio's success was not solely dependent on the AI model; it also benefited from careful risk management and rigorous backtesting. The model underwent extensive backtesting, using historical data to simulate its performance under various market conditions. This rigorous validation process helped to identify potential weaknesses and refine the model's parameters before deploying it to manage real capital. Moreover, the investment firm implemented a robust risk management framework, incorporating stringent position limits, diversification strategies, and regular portfolio reviews. This multi-layered approach ensured that the AI model's potential benefits were realized without exposing the portfolio to undue risk.

Beyond traditional financial metrics, the firm also monitored the AI model's explainability. While deep learning models are often considered "black boxes," the firm employed techniques to improve the interpretability of the model's decisions. This involved analyzing the model's feature importance, providing insights into which factors had the greatest impact on its predictions. This not only aided in understanding the model's underlying logic but also facilitated compliance with regulatory requirements and ensured transparency for investors. This transparency was crucial in building investor confidence and attracting capital. Regular reports detailing the model's performance, its underlying logic, and the rationale behind investment decisions were provided to clients.

The portfolio's composition was remarkably dynamic, adapting to market shifts and opportunities. During periods of uncertainty, it often held larger positions in defensive assets like government bonds and high-quality corporate debt. As confidence returned and growth prospects improved, it would progressively increase its exposure to equities, particularly companies identified by the AI model as having high growth potential. This responsiveness enabled the portfolio to capture both defensive and offensive opportunities, contributing to its consistent outperformance. The model's capacity to adapt was tested and proven during periods of significant market turbulence, such as the initial COVID-19 market crash. While most portfolios suffered significant losses, this AI-driven portfolio experienced a relatively muted decline and subsequently recovered swiftly.

The AI model's ability to analyze alternative data sources proved particularly valuable. For example, the model's analysis of satellite imagery of retail parking lots enabled it to predict sales trends for certain retail companies more accurately than traditional methods. This alternative data provided an edge in identifying underperforming or outperforming companies, allowing the portfolio to capitalize on market inefficiencies. The integration of social media sentiment analysis further enhanced the model's predictive capabilities. By analyzing the tone and sentiment expressed in social media posts related to specific companies, the model could anticipate shifts in market sentiment more rapidly than relying solely on traditional news sources.

However, the successful implementation of this AI-driven strategy also faced challenges. Maintaining and updating the AI model required significant computational resources and ongoing expertise in both AI and finance. The constant need for data updates, model retraining, and algorithm refinement presented a considerable operational challenge. The firm had to make substantial investments in both technology and personnel to support the AI strategy. Furthermore, regulatory compliance and the need for transparency in AI-driven investment decision-making presented ongoing hurdles. The firm had to carefully navigate the regulatory landscape, ensuring compliance with all applicable laws and regulations. This required close collaboration with legal and compliance experts.

The case study of this AI-driven portfolio demonstrates the significant potential of integrating advanced AI techniques into investment management. However, it also underscores the complexities and challenges involved in successfully deploying such strategies. It's a testament to the fact that AI is not a magic bullet; it's a powerful tool that requires careful implementation, rigorous testing, ongoing maintenance, and a deep understanding of both AI and financial markets to be truly effective. The combination of cutting-edge technology and seasoned financial expertise was crucial in achieving the portfolio's impressive results. While AI enhanced the speed and accuracy of decision-making, human oversight remained essential in managing risk, ensuring regulatory compliance, and interpreting the model's outputs. This symbiotic relationship between human intelligence and artificial intelligence stands as a key takeaway from this successful case study. The future of investment management likely lies in this collaborative approach.

The previous case study showcased the consistent outperformance of an AI-driven portfolio during relatively stable market conditions. However, the true mettle of any investment strategy is revealed during periods of crisis. This second case study examines the performance of a different AI-driven investment strategy during the acute phase of the 2022 global inflation crisis and subsequent interest rate hikes. This period presented a unique challenge due to the unexpected and rapid shift in macroeconomic conditions, dramatically altering the typical correlations between asset classes and significantly increasing market volatility.

This specific investment firm, unlike the previous one, focused on a multi-asset class approach. Their proprietary AI model, based on a reinforcement learning architecture, wasn't solely focused on predicting market direction but rather on optimizing portfolio allocations across a diverse range of asset classes including equities, bonds, commodities, and alternative investments like real estate investment trusts (REITs) and private equity (through carefully selected funds). The model's training data encompassed a vast array of inputs – traditional macroeconomic data such as inflation rates, interest rate forecasts, GDP growth estimates, and unemployment figures; sentiment data gleaned from news articles, social media, and financial analyst reports via natural language processing (NLP); and alternative data sources such as supply chain indices, credit default swap spreads, and consumer spending data derived from point-of-sale systems.

The model's distinguishing feature was its dynamic risk management capabilities. Recognizing that traditional Value-at-Risk (VaR) models proved insufficient in such a rapidly changing environment, this AI system employed a novel approach combining advanced simulation techniques with real-time market data. Instead of relying solely on historical volatility, the model leveraged Monte Carlo simulations to project a wide range of potential future scenarios, considering different interest rate paths, inflation scenarios, and geopolitical events. Each scenario was assigned a probability based on the AI's analysis of the current market landscape. This allowed for a highly adaptable strategy, altering asset allocation based on the probabilities of these different scenarios playing out.

For instance, as inflation began its rapid ascent in early 2022, the model downshifted its allocation to equities, particularly those in growth sectors which are typically more sensitive to interest rate increases. The algorithm correctly anticipated the downward pressure on growth stock valuations.

Simultaneously, it increased allocations to inflation-hedged assets such as commodities (particularly gold and energy) and inflation-protected securities (TIPS). This proactive response prevented significant losses that many traditional portfolios experienced during this inflationary surge.

The AI's ability to process and interpret diverse data sources proved invaluable. While many analysts focused on headline inflation numbers, the AI model also considered the underlying drivers of inflation, such as supply chain bottlenecks and energy prices. By analyzing alternative data sources, such as real-time supply chain indices, it could assess the severity and persistence of these bottlenecks and adjust its portfolio accordingly. This nuanced understanding enabled more accurate predictions than simpler models relying solely on aggregate inflation figures.

Further enhancing its predictive power, the model incorporated sophisticated sentiment analysis from various sources. It could discern differences in sentiment between retail investors, institutional investors, and financial professionals. This allowed for a more granular understanding of market dynamics, helping to anticipate shifts in investor behavior before they manifested in actual price movements. For example, the algorithm detected growing pessimism among retail investors regarding the technology sector well before this sentiment was reflected in broader market indices. This early warning system allowed the firm to reduce its exposure to technology stocks, mitigating potential losses.

The model's capacity to rapidly adjust its allocations was crucial in navigating the frequent and sometimes unpredictable shifts in market conditions. As central banks began aggressively raising interest rates to combat inflation, the model recognized that bond yields were becoming more attractive. It strategically shifted a portion of the portfolio into higher-yielding bonds, capitalizing on the rising interest rate environment. The speed and precision of these adjustments, often executed within minutes of receiving updated market data, were impossible to replicate manually.

However, the implementation wasn't without its challenges. The firm faced significant computational costs in running the sophisticated simulations and processing large volumes of data. Furthermore, maintaining and updating the model required a team of data scientists, financial engineers, and investment professionals with a high level of expertise. This human capital investment was crucial for ensuring the accuracy and reliability of the AI model. The explainability of the model's decisions was also a critical consideration. The firm employed advanced techniques like SHAP (SHapley Additive exPlanations) values to offer insights into the model's decision-making process, ensuring transparency for both internal stakeholders and regulatory bodies.

The firm also faced regulatory hurdles, needing to demonstrate to regulators the robustness and transparency of their AI system. This necessitated detailed documentation of the model's architecture, training data, and decision-making process. Regular audits and rigorous testing were crucial in ensuring ongoing compliance.

Ultimately, the AI-driven strategy proved highly effective in navigating the 2022 market crisis. While other portfolios suffered significant losses, this portfolio demonstrated resilience, significantly outperforming its benchmarks. The strategy's success wasn't simply due to the AI model's predictive capabilities but also resulted from the firm's integrated approach, combining cutting-edge technology with traditional investment principles and experienced human oversight. This case study underscores the importance of a hybrid model combining human expertise and AI's analytical power for effective investment management in a volatile market environment.

The comparative analysis between this case study and the previous one highlights the versatility of AI in investment management. While the first case study demonstrated consistent outperformance in a relatively stable market through optimization and predictive analytics, the second case study showcases AI's ability to navigate sudden market shocks through dynamic risk management and rapid adaptation. Both case studies, however, underscore the importance of a well-defined investment strategy, rigorous backtesting, and continuous monitoring of the AI models, as well as a keen awareness of the ethical and regulatory considerations associated with AI-driven investing. The future of successful investment management will undoubtedly involve a seamless integration of advanced AI technologies with human expertise and sound financial principles. This collaboration will be crucial in maximizing returns while effectively mitigating risks in an ever-evolving and increasingly complex financial landscape. The success stories of these AI-driven investment strategies highlight not only the potential but also the crucial responsibilities associated with developing, deploying and utilizing AI in the financial world. The ethical considerations of transparency, explainability, and risk management will remain critical considerations for the future of AI in finance, ensuring that technological advancements serve to improve, not endanger, the financial stability of individuals and markets alike. The ongoing evolution of AI necessitates a constant review of regulatory frameworks and ethical guidelines, ensuring that these powerful tools are used responsibly and for the benefit of all stakeholders.

Private equity investments, by their very nature, involve a higher degree of uncertainty compared to publicly traded securities. The information asymmetry between the investor and the target company is often significant, making thorough due diligence absolutely crucial. Traditional due diligence relies heavily on manual review of financial statements, legal documents, and market research, a process that can be time-consuming, expensive, and prone to human error. The introduction of AI offers a transformative potential to enhance this critical stage of the investment process. This case study focuses on a large private equity firm, let's call it "Apex Capital," that integrated AI-powered tools into its due diligence workflow.

Apex Capital's strategy centered on augmenting, not replacing, its human analysts. The firm recognized the inherent limitations of relying solely on AI and maintained a strong emphasis on human expertise and judgment. Their AI system, built in partnership with a leading data science firm, focused on several key areas within the due diligence process.

One critical application involved the analysis of vast amounts of unstructured data. Traditional methods struggle to efficiently process information scattered across news articles, social media posts, regulatory filings, and industry publications. Apex Capital's AI leveraged natural language processing (NLP) and machine learning (ML) techniques to sift through this data, identifying key trends, risks, and opportunities that might otherwise be missed. For instance, in analyzing a potential acquisition target in the renewable energy sector, the AI system identified a subtle shift in public sentiment regarding government subsidies for solar energy projects, a development that was not yet widely reflected in traditional financial analysis. This early warning signal prompted the due diligence team to delve deeper into the potential impact on the target company's future profitability. This additional scrutiny allowed them to negotiate a more favorable price than they would have initially been willing to pay, taking into account potential future government policy changes.

Furthermore, the AI system excelled at identifying anomalies and inconsistencies within the target company's financial statements. Using advanced anomaly detection algorithms, the system could pinpoint unusual fluctuations in revenue streams, expenses, or accounting practices that might indicate fraudulent activity or poor management. In one instance, the AI flagged a discrepancy in the target company's inventory valuation, prompting the human analysts to conduct a more thorough investigation. This investigation revealed an overstatement of inventory value by approximately 15%, a material misstatement that significantly impacted the perceived financial health of the company. This discovery allowed Apex Capital to either renegotiate the deal terms or walk away from the investment entirely, preventing a potentially substantial financial loss.

The AI's ability to rapidly process and correlate information from diverse sources was a significant advantage. It could link data points across various databases, identify hidden connections, and generate insightful visualizations that significantly aided the human analysts in their decision-making process. For example, the system connected data points revealing a decline in customer satisfaction scores with an increase in employee turnover, indicating potential operational problems. This insight would have been difficult to uncover through traditional analysis, yet it provided critical context for the analysts assessing the long-term sustainability and profitability of the target company.

The system also contributed to a more efficient use of human resources. By automating repetitive and time-consuming tasks, such as data entry and preliminary financial analysis, the AI freed up human analysts to focus on higher-level tasks such as evaluating strategic fit, assessing management teams, and conducting in-depth industry research. This optimization of human resources allowed Apex Capital to conduct due diligence more rapidly and thoroughly, accelerating its investment cycle and improving the overall efficiency of its investment process.

Beyond financial analysis, the AI system was also applied to assess environmental, social, and governance (ESG) factors. The system incorporated ESG ratings from third-party providers, as well as news articles and social media sentiment to form a comprehensive ESG profile of each potential investment target. This was particularly useful in helping Apex Capital to assess potential reputational risks associated with investment choices. In one notable case, the AI identified a potential ethical concern relating to the labor practices of a target company's supplier, which was not immediately apparent from traditional sources. This discovery facilitated a more robust dialogue with the target company, leading to concrete steps towards improvement of these practices.

Despite the significant benefits, Apex Capital faced challenges in implementing AI into its due diligence process. The integration of the AI system required substantial investment in infrastructure, data scientists, and training for existing employees. Furthermore, ensuring the accuracy, reliability, and explainability of the AI model was crucial. The firm addressed this by implementing rigorous testing protocols, conducting regular audits, and employing techniques like SHAP values to explain the AI's decision-making process, enhancing transparency for all stakeholders.

The firm also had to consider data privacy and security concerns. The AI system processed highly sensitive data, requiring robust security measures to prevent unauthorized access and data breaches. Compliance with relevant data protection regulations was a paramount concern throughout the implementation and use of the AI system.

The success of Apex Capital's AI-enhanced due diligence process is undeniable. The firm demonstrated a demonstrable reduction in the time and cost associated with due diligence while significantly improving the quality and depth of its analysis. The integration of AI led to the identification of several critical risk factors and opportunities that would have been otherwise missed. Moreover, it improved the overall efficiency of the investment process, contributing to a faster and more informed decision-making process. The enhanced due diligence capabilities significantly contributed to Apex Capital's ability to identify high-quality investments, negotiate favorable terms, and ultimately achieve superior returns. This case study demonstrates that AI isn't just a futuristic concept in private equity—it's a transformative technology already reshaping the industry, enhancing both efficiency and the overall quality of investment decisions. The future of due diligence will undoubtedly be driven by a symbiotic relationship between human expertise and the analytical power of artificial intelligence.

This case study delves into the intricacies of a mean reversion algorithmic trading strategy, focusing on its application in the highly volatile cryptocurrency market. The strategy, developed by a quantitative trading firm we'll refer to as "QuantAlpha," aimed to capitalize on the frequent price fluctuations inherent in crypto assets. The core principle was simple: buy when prices fall below a calculated mean and sell when they rise above it, profiting from the expected reversion to the average. However, the implementation was far from simplistic, requiring sophisticated AI algorithms and rigorous risk management protocols.

QuantAlpha's strategy relied on a multi-layered approach. The first layer involved the selection of cryptocurrencies for trading. Instead of a broad-market approach, QuantAlpha focused on a curated basket of established cryptocurrencies with significant trading volume and liquidity. This selection process utilized machine learning algorithms that analyzed historical price data, trading volume, market capitalization, and social media sentiment. The algorithms identified cryptocurrencies that exhibited a statistically significant tendency towards mean reversion within specific timeframes, while simultaneously filtering out those deemed too volatile or illiquid for the strategy's risk tolerance.

The second layer involved the development of sophisticated mean reversion models. These models went beyond simple moving averages, incorporating more advanced statistical techniques such as Kalman filtering and exponential smoothing. Kalman filtering proved particularly useful in handling noisy data, a common characteristic of cryptocurrency markets prone to sudden price spikes and crashes. The model dynamically adjusted to changing market conditions, adapting its mean reversion thresholds based on real-time data streams. This adaptive capability was crucial in mitigating the risks associated with market regime shifts. The algorithm wasn't static; it learned and adjusted based on its performance and the evolving market dynamics.

Furthermore, the model incorporated volatility clustering into its calculations. Volatility clustering, a phenomenon where periods of high volatility are followed by further high volatility, was accounted for to adjust trading signals and position sizing accordingly. During periods of heightened volatility, the algorithm would reduce its trading frequency and position size, limiting exposure to potentially devastating price swings. Conversely, during quieter periods, it would increase its trading activity. This dynamic risk management approach was essential for long-term strategy sustainability.

The third layer encompassed risk management. QuantAlpha implemented stringent stop-loss orders to limit potential losses on individual trades. These stop-loss orders were dynamically adjusted based on the volatility of each cryptocurrency and the overall market conditions. The algorithm also incorporated position sizing mechanisms to prevent excessive exposure to any single cryptocurrency. This diversification strategy aimed to reduce portfolio risk and protect against large, unexpected price movements in any individual asset. Furthermore, the system included a "circuit breaker" mechanism that would temporarily suspend trading activity if certain market thresholds were breached, such as a sudden and sharp drop in the overall cryptocurrency market.

The backtesting phase of the strategy was extensive and rigorous. QuantAlpha used several years of historical cryptocurrency data to simulate the performance of the trading algorithm under various market conditions. This backtesting not only validated the core mean reversion concept but also identified potential weaknesses and areas for improvement. The backtesting also explored various parameter settings for the algorithms, optimizing them for maximum risk-adjusted returns. Sensitivity analysis was extensively performed to understand the impact of various input variables on the overall strategy performance.

The live implementation of the strategy involved careful monitoring and ongoing optimization. QuantAlpha employed a team of quantitative analysts who continuously monitored the algorithm's performance, identifying potential issues and making adjustments as needed. The system also generated comprehensive performance reports, providing detailed insights into the strategy's profitability, risk profile, and trading activity. These reports were crucial for identifying potential areas of improvement and informing strategic decision-making.

Despite the rigorous development and testing, QuantAlpha encountered significant challenges during the live trading phase. One major challenge was the inherent unpredictability of the cryptocurrency market. Unexpected news events, regulatory changes, and shifts in investor sentiment could trigger significant price swings that defied the strategy's underlying assumptions. For example, a sudden regulatory crackdown on a particular cryptocurrency could result in a massive price drop, overwhelming the strategy's stop-loss orders. The algorithm, while adaptive, wasn't immune to unforeseen black swan events.

Another challenge was the phenomenon of "flash crashes," where prices experience extremely rapid and dramatic declines. These flash crashes could trigger the algorithm's stop-loss orders, resulting in losses even if the price eventually recovered. This highlighted the limitations of a purely quantitative approach, emphasizing the importance of considering qualitative factors alongside quantitative data.

Moreover, the strategy's performance was susceptible to market regime shifts. During periods of high volatility and strong trending behavior, the mean reversion strategy could underperform, as prices didn't always revert to their historical means. This necessitated adjustments to the algorithm's parameters and potentially even a shift in strategy during such periods. The team had to continually refine the algorithms to better adapt to different market conditions and to account for evolving cryptocurrency market dynamics.

Data quality also proved to be a significant challenge. The cryptocurrency market is characterized by relatively low regulatory oversight and a lack of standardized data sources. This resulted in inconsistent and unreliable data, which could impact the accuracy of the algorithm's predictions and lead to suboptimal trading decisions. Data cleansing and validation became a significant ongoing task, requiring constant monitoring and refinement of data sources.

In spite of these challenges, QuantAlpha's algorithmic trading strategy demonstrated a positive risk-adjusted return over the long term. The strategy's adaptive nature and robust risk management protocols helped to mitigate potential losses during periods of high volatility. The continuous monitoring and optimization of the algorithm proved crucial to its success. However, the experience underscored the importance of a blended approach, combining the power of AI and quantitative analysis with human oversight and intuition. The success of algorithmic trading in the cryptocurrency market, therefore, lies not just in the sophistication of the algorithms themselves but in the ability to manage the inherent risks and adapt to the constantly evolving dynamics of this unique and volatile market. The case demonstrates that while AI can significantly enhance trading capabilities, human judgment and adaptation remain irreplaceable elements for consistent success. The future of algorithmic trading likely lies in the intelligent collaboration between humans and machines, leveraging the strengths of each to navigate the complex and ever-changing landscape of financial markets.

This case study focuses on the application of sentiment analysis to predict market movements, specifically a significant downturn in the technology sector. The prediction, made several weeks in advance, allowed a hedge fund, which we will refer to as "Apex Capital," to successfully mitigate substantial losses and even profit from the subsequent market correction. Unlike the mean reversion strategy employed by QuantAlpha, this approach utilized a fundamentally different methodology, relying on the analysis of unstructured data to gauge investor sentiment and anticipate market trends.

Apex Capital's strategy centered on the hypothesis that collective investor sentiment, as reflected in various online platforms, could serve as a leading indicator of market movements. The team believed that shifts in overall sentiment, from optimistic to pessimistic, could precede actual price declines, offering a valuable window of opportunity for proactive risk management. This approach required the development of sophisticated AI algorithms capable of processing vast amounts of unstructured data, including news articles, social media posts, financial blogs, and online forums.

The data acquisition phase was a significant undertaking. Apex Capital leveraged a combination of web scraping techniques and partnerships with specialized data providers to gather a comprehensive dataset encompassing a wide range of online sources. The data was collected in real-time, ensuring that the analysis reflected the most current investor sentiment. This constant data stream was crucial, as the speed at which sentiment shifts can impact markets necessitates immediate responsiveness. The challenge was not only collecting the data but also ensuring its quality and relevance, a process that required meticulous filtering and cleaning. Irrelevant or low-quality data could have easily skewed the results, leading to inaccurate predictions. Therefore, rigorous data quality control measures were established and continuously monitored.

The core of Apex Capital's AI-driven sentiment analysis system relied on natural language processing (NLP) techniques. These algorithms were designed to identify and categorize the emotional tone expressed in the collected textual data. The system was trained on a massive corpus of previously labeled data, enabling it to distinguish between positive, negative, and neutral sentiments. This involved the use of advanced machine learning models, specifically recurrent neural networks (RNNs) and long short-term memory (LSTM) networks, which are particularly well-suited for processing sequential data like text. These models were capable of not only identifying the sentiment of individual sentences or posts but also understanding the broader context and recognizing nuanced expressions of sentiment.

Beyond simple sentiment classification, Apex Capital incorporated more sophisticated NLP techniques, such as topic modeling and named entity recognition. Topic modeling helped to identify the key themes and subjects driving investor sentiment, providing valuable context for the overall sentiment score. For example, identifying a surge in negative sentiment related to a specific company's earnings report allowed for a more targeted and precise assessment of risk. Named entity recognition helped to identify key players and entities mentioned in the data, enabling the algorithm to track sentiment shifts associated with specific companies, industries, or even individuals.

The system was not simply aggregating sentiment scores; it was designed to detect subtle shifts and patterns within the data. A significant component of the AI's function involved anomaly detection, flagging sudden and unexpected shifts in sentiment that could signal impending market disruptions. This required the development of sophisticated statistical models capable of differentiating between normal fluctuations and significant deviations from established baselines. These models incorporated time series analysis techniques to account for the temporal dependencies inherent in sentiment data. The system was continually learning and adapting, refining its predictive capabilities through continuous feedback loops and retraining using updated data.

Once the sentiment data was analyzed, the system generated a comprehensive sentiment score reflecting the overall market sentiment across the various data sources. This score was then integrated with a range of quantitative factors, such as economic indicators, market volatility, and historical price trends. This blended approach, combining qualitative sentiment data with quantitative market data, allowed for a more robust and nuanced assessment of market risk. The integration of various data types was achieved through advanced statistical methods, enabling the model to weight different factors according to their relative importance and predictive power.

The prediction of the technology sector downturn emerged from a prolonged period of increasingly negative sentiment detected by the AI system. Over several weeks, the system consistently flagged a sharp increase in negative sentiment across multiple data sources, coupled with a decrease in positive sentiment. The thematic analysis identified several key drivers, including concerns about rising interest rates, slowing economic growth, and overvaluation within the technology sector. This convergence of negative sentiment across multiple channels, alongside corroborating quantitative market signals, triggered the algorithm to issue a strong warning signal indicating a high probability of an impending market correction.

Apex Capital responded swiftly to the AI's prediction. They initiated a series of risk mitigation strategies, including reducing their exposure to the technology sector, increasing their cash position, and employing hedging techniques to protect against further losses. These proactive measures were instrumental in limiting their losses during the subsequent market downturn. In fact, by leveraging the insights provided by the AI, Apex Capital was not only able to avoid significant losses but also managed to capitalize on the downturn by strategically investing in undervalued assets during the correction.

The success of this case study underscores the transformative potential of AI-driven sentiment analysis in financial markets. The ability to harness the power of unstructured data, analyzing investor sentiment in real-time, provides a significant advantage in anticipating market trends and managing risk. However, it also highlights the need for a sophisticated and robust AI system, capable of processing massive datasets, handling noisy data, and integrating qualitative information with quantitative factors. The success wasn't just about the algorithm; it was about the careful integration of the AI with human expertise, ensuring that the output of the AI was interpreted and acted upon effectively. The combination of technology and human judgment proved crucial to effectively utilizing the insights.

It is important to acknowledge that even the most sophisticated AI systems are not infallible. Sentiment analysis is inherently complex, and unexpected events can always disrupt market predictions. While Apex Capital's system demonstrated impressive accuracy in this specific case, it's crucial to acknowledge that this success does not guarantee future performance. The system is continuously being refined and improved to enhance its accuracy and adaptability. The team also conducts regular stress tests and sensitivity analysis to understand the potential impact of various factors and to identify areas for improvement.

Moreover, the ethical considerations surrounding the use of sentiment analysis in financial markets should not be overlooked. The potential for bias in the data, algorithmic discrimination, and market manipulation needs to be carefully addressed. Robust mechanisms for data validation, bias detection, and responsible AI implementation are essential to ensuring that these technologies are used ethically and responsibly. Transparency and accountability are paramount in ensuring that the use of AI in finance benefits all market participants. Apex Capital maintains a rigorous internal ethics framework guiding the development and deployment of their AI systems, prioritizing fairness, transparency, and the avoidance of potentially harmful applications.

Finally, the case study serves as a compelling example of the evolution of AI-driven investing. As AI technologies continue to advance, their role in financial markets will undoubtedly expand, offering new opportunities for enhanced risk management, improved investment strategies, and more efficient market operations. The future of finance will likely witness a more significant integration of AI, however, responsible development and ethical considerations must remain at the forefront, ensuring a fair and equitable financial ecosystem. The successful application of AI hinges not only on technical proficiency but also on ethical awareness and a commitment to responsible innovation.

The successful application of AI in investment strategies, as demonstrated by Apex Capital, necessitates the careful selection of appropriate tools and platforms. This is not a one-size-fits-all scenario; the optimal choice depends heavily on individual needs, investment strategies, and technical expertise. Choosing the wrong tool can lead to wasted resources, inaccurate results, and ultimately, poor investment decisions. This section provides a practical framework for navigating the landscape of AI investment tools, emphasizing key considerations and offering specific examples.

One of the first hurdles is defining your specific needs. What are your primary investment goals? Are you focused on algorithmic trading, portfolio optimization, risk management, or sentiment analysis, as was the case with Apex Capital? Understanding your objectives will significantly narrow down the field of potential AI tools. For instance, if your focus is on high-frequency trading, you'll require tools with extremely low latency and robust execution capabilities. Conversely, if your focus is on long-term value investing, your needs may be better served by tools that excel at fundamental analysis and predictive modeling.

Cost is another critical consideration. AI tools and platforms vary significantly in price, ranging from free open-source options to enterprise-grade solutions costing tens of thousands of dollars per year. The cost must be weighed against the potential benefits. While expensive platforms often offer more advanced functionalities and better support, free or low-cost options can be suitable for smaller operations or those in the early stages of AI adoption. Careful evaluation of the features offered at each price point is crucial to ensure value for money. A thorough cost-benefit analysis, considering both upfront costs and ongoing maintenance expenses, is essential.

Functionality is paramount. Different AI tools offer diverse functionalities, encompassing data acquisition, data cleaning and preprocessing, model building, model training, backtesting, and deployment. You need to assess which functionalities are essential to your investment strategy and which can be outsourced or handled manually. Some tools excel at specific tasks, such as natural language processing (NLP) for sentiment analysis, while others offer more comprehensive capabilities, encompassing various aspects of the AI investment lifecycle. A comprehensive assessment of the tool's feature set against your requirements is crucial for a well-informed decision.

Ease of use should not be overlooked. While advanced AI tools often require significant technical expertise, user-friendliness is a significant factor, especially for those with limited programming skills or data science backgrounds. Some platforms offer user-friendly interfaces and pre-built models that require minimal coding, making them accessible to a broader range of users. Others demand a high level of technical proficiency and extensive customization. Consider your team's technical skills and choose a tool that aligns with their capabilities. Training and support resources offered by the vendor should also be factored into the evaluation.

Beyond these fundamental considerations, several other factors warrant attention. Data integration capabilities are crucial, as AI tools often need to interact with various data sources, including market data providers, news feeds, and social media platforms. Compatibility with existing infrastructure and systems is also essential, avoiding the need for extensive system overhauls. Scalability is another key factor, ensuring the tool can handle growing data volumes and increasing computational demands as your needs evolve. Finally, security and compliance are paramount, particularly in the financial industry, where data protection and regulatory compliance are stringent. Selecting a tool with robust security measures and adherence to relevant regulations is non-negotiable.

Let's examine some specific examples of AI tools used in investment management. For algorithmic trading, platforms like MetaTrader 5, NinjaTrader, and TradingView offer varying degrees of automation and customization. MetaTrader 5 is a popular choice for its extensive capabilities and large community support. However, it requires considerable technical expertise. NinjaTrader provides a more user-friendly interface, while TradingView focuses on charting and technical analysis, integrating AI-powered indicators.

For portfolio optimization, tools like PortfolioVisualizer and Riskalyze provide quantitative analysis and scenario planning. PortfolioVisualizer allows for backtesting various portfolio strategies, while Riskalyze focuses on risk management and aligning portfolios with client risk tolerance. These tools are often less technically demanding than algorithmic trading platforms, making them accessible to a wider audience.

In the realm of sentiment analysis, several options exist. Platforms like Aylien and MonkeyLearn offer NLP capabilities, allowing for the analysis of news articles, social media posts, and other unstructured data. These tools generally require some programming skills, although some offer pre-built models and APIs for easier integration. The choice will depend on your comfort level with coding and the sophistication of the analysis required. More specialized platforms may offer pre-trained models specific to financial markets, potentially offering better accuracy and less need for customization.

For broader machine learning tasks, platforms like Google Cloud AI Platform, Amazon SageMaker, and Microsoft Azure Machine Learning provide comprehensive environments for building, training, and deploying machine learning models. These are typically aimed at experienced data scientists and require significant technical expertise. Their versatility allows for the development of custom models tailored to specific investment strategies, but this comes with a steeper learning curve and higher resource requirements.

Open-source libraries like TensorFlow and PyTorch provide the building blocks for creating custom AI models, offering maximum flexibility but demanding significant programming expertise. These are ideal for highly specialized tasks where pre-built solutions are inadequate, but they require considerable technical skill and time investment.

The selection of the right AI tools is an iterative process. It's advisable to start with a clear definition of your needs, followed by a thorough evaluation of available tools based on cost, functionality, ease of use, data integration capabilities, scalability, security, and compliance requirements. Don't hesitate to experiment with trial versions or free plans before committing to a paid subscription. Consider the technical capabilities of your team and choose a tool that aligns with their skill set. A phased approach, starting with simpler tools and gradually incorporating more advanced features as your expertise grows, can be a wise strategy. Remember that the most sophisticated tool is useless if it isn't understood and used effectively. The human element remains crucial in interpreting the AI's output and making informed investment decisions. Continuous learning and adaptation are essential in the ever-evolving world of AI-driven finance. The effective integration of technology and human expertise remains the key to success.

Integrating AI into an established investment process requires a methodical approach, moving beyond simply acquiring new tools to fundamentally altering workflows and decision-making. It's not merely about adding AI; it's about transforming how you work. This involves careful planning, phased implementation, and continuous monitoring and adaptation. A rushed or poorly planned integration can lead to disruptions, inaccuracies, and ultimately, diminished returns.

The first critical step is a thorough audit of your current investment process. Document every step, from data acquisition and analysis to portfolio construction and risk management. Identify bottlenecks, inefficiencies, and areas where AI could potentially add value. This detailed understanding of your existing workflow forms the foundation for a successful AI integration. For instance, if your process relies heavily on manual data entry and analysis, automating these tasks with AI could significantly improve efficiency and reduce human error. Similarly, if your risk management relies solely on historical data, incorporating AI-driven predictive modeling could provide a more comprehensive assessment of future risks.

Once you've thoroughly analyzed your existing process, the next step is to identify specific AI tools and techniques that address your identified needs. This requires matching the capabilities of available AI tools to your specific challenges and opportunities. Remember, AI isn't a magic bullet; the appropriate application is crucial. Consider different AI capabilities, such as predictive modeling for forecasting market trends, natural language processing (NLP) for sentiment analysis from news and social media, or machine learning algorithms for portfolio optimization. Do not simply choose the most advanced tool; choose the tool that best fits your workflow and expertise.

The integration itself should be a phased process. Starting with a pilot project involving a relatively small segment of your investment process allows you to test the AI's performance and identify potential issues before a full-scale deployment. This could involve integrating an AI-powered sentiment analysis tool to supplement your existing news monitoring process or using an AI-driven portfolio optimization tool to refine a small portion of your portfolio. This phased approach minimizes risk and allows for iterative improvements.

Data integration is paramount. AI tools require vast amounts of clean, consistent data to function effectively. This often involves integrating data from multiple sources, including internal databases, market data providers, and external news sources. Ensuring data consistency and quality is vital to avoid inaccuracies in AI predictions and recommendations. Implementing robust data cleaning and preprocessing procedures is crucial. This might involve standardizing data formats, handling missing values, and identifying and removing outliers. Data quality is paramount. Garbage in, garbage out is a truism that holds especially true for AI-driven investment strategies. The investment in data preparation and management will be crucial to success.

Following the pilot phase, gradually expand the AI's role in your investment process. As confidence in the AI's performance grows, you can increase its involvement in decision-making, but this expansion should be monitored carefully. Continuous evaluation of the AI's output is essential, comparing its performance against existing methods and identifying any discrepancies or anomalies. This requires a dedicated team responsible for monitoring the AI's performance and making necessary adjustments. This may involve retraining the AI model with new data, fine-tuning its parameters, or adjusting its integration with other elements of your workflow.

Moreover, effective integration requires changes in your team's approach. This means providing adequate training to your staff on using and interpreting the output of AI tools. You may need to invest in training programs to enhance your team's understanding of AI and its application in investment management. This is particularly important if your team has limited experience with AI technologies. Furthermore, fostering a culture of continuous learning and improvement within your organization is critical to maintaining the competitiveness of your investment process in the rapidly evolving landscape of AI-driven finance.

Consider the ethical implications. AI in finance introduces complexities regarding bias, transparency, and explainability. Ensure your AI models are designed to mitigate potential biases and that their decision-making processes are transparent and understandable. This is crucial for maintaining trust and compliance with relevant regulations. Addressing these ethical considerations proactively is not merely a compliance issue; it is essential for maintaining the integrity of your investment strategy and building trust with clients.

In a practical example, consider a hedge fund manager currently using spreadsheet-based methods for portfolio optimization. To integrate AI, the manager could begin by importing historical portfolio performance data into a machine learning platform like Python with scikit-learn. After data cleaning and pre-processing, they might train a simple regression model to predict portfolio returns based on various market factors. This model could then suggest optimal asset allocations within a small, controlled segment of the portfolio. Over time, as confidence in the model increases, its usage can be expanded to encompass a larger portion of the portfolio. The manager will need to closely monitor the model's performance, comparing it to previous portfolio strategies and iteratively refining the model based on performance analysis and feedback. This allows a controlled and measured transition of the workflow.

Another example could involve a wealth management firm that currently relies on human analysts to assess market sentiment. They might introduce an AI-powered natural language processing tool to analyze news articles and social media posts, supplementing human analysis. Initially, the AI's output could be treated as a supplementary data point for the human analysts, allowing them to compare the AI's assessment with their own. Gradually, as the AI's accuracy improves, it can play a larger role in informing investment decisions. This approach ensures a smooth integration of AI while preserving human oversight and expertise.

The integration of AI into investment processes is an ongoing journey, not a destination. Regularly update and retrain your AI models with new data to maintain their accuracy and relevance. The financial markets are dynamic and constantly evolving, and your AI models need to adapt to these changes. Regular model evaluation, retraining, and potential updates to underlying algorithms are integral to long-term success. Regular monitoring of performance, alongside rigorous testing and validation, are paramount for maintaining accuracy and ensuring the AI continues to deliver value. The key is continuous improvement and adaptation to changes in the market.

Successful AI integration goes beyond technological implementation; it fundamentally redefines the roles and responsibilities within an investment team. Human expertise becomes increasingly focused on interpreting AI-generated insights, managing model risk, formulating strategic investment decisions, and establishing ethical guidelines and regulatory compliance. The balance between human judgment and AI-driven analysis represents the ultimate power of AI in finance. The human element remains crucial in interpreting the output and making informed decisions, ensuring the responsible and ethical application of this powerful technology.

The successful integration of AI into investment strategies hinges critically on robust data management and unwavering security protocols. The sheer volume and sensitivity of financial data involved necessitate a meticulously planned approach to storage, access control, and cybersecurity, mitigating risks and ensuring compliance with regulatory requirements. Failing to prioritize these aspects can lead to significant financial losses, reputational damage, and legal repercussions.

First, let's address data storage. AI models thrive on data; the more comprehensive and higher-quality the data, the more accurate and effective the model's predictions. However, managing this data requires a sophisticated system. Simple spreadsheet solutions are woefully inadequate. Instead, consider cloud-based data warehousing solutions that provide scalability, redundancy, and accessibility. These platforms, such as AWS S3, Azure Blob Storage, or Google Cloud Storage, offer secure and cost-effective ways to manage massive datasets. They typically include features for data versioning, allowing you to revert to previous data states if necessary, a crucial safeguard against accidental data corruption. Furthermore, these services usually incorporate robust encryption protocols to protect data at rest and in transit.

Beyond the choice of platform, the structure of your data is crucial. Adopting a well-defined data schema, with clear data types and relationships between different data points, enhances data quality and simplifies analysis. This schema should be meticulously documented, making it easily understandable by both human analysts and the AI algorithms. Consistent data formatting across all sources is paramount; inconsistent formats will introduce errors and reduce the reliability of your AI's predictions.

Data access control is another critical aspect. Not all members of your investment team need access to all data. Implementing a granular access control system, typically through role-based access control (RBAC), is essential. RBAC allows you to assign specific permissions to different users based on their roles within the organization. For example, senior portfolio managers might have full access to all data, while junior analysts might only have access to specific datasets relevant to their tasks. This minimizes the risk of unauthorized data access and strengthens the overall security posture. Regular audits of user access rights should be conducted to identify and rectify any potential vulnerabilities. A robust logging system that tracks all data access events is crucial for both security and compliance.

The security of your data infrastructure is paramount. Cybersecurity threats are ever-present, and the financial industry is a prime target for malicious actors. Implementing a multi-layered security approach is crucial. This includes robust firewalls, intrusion detection systems (IDS), and intrusion prevention systems (IPS) to protect your data from external attacks. Regular security audits and penetration testing by external security specialists should be incorporated into your data management strategy to identify and address vulnerabilities before they can be exploited. Employee training on cybersecurity best practices is also vital, educating employees about phishing scams, malware, and other potential threats. Furthermore, adherence to industry best practices and compliance with relevant regulations, such as GDPR or CCPA, are essential to protect sensitive client data.

Data encryption is a fundamental aspect of securing financial data. Data encryption transforms data into an unreadable format, protecting it from unauthorized access even if a security breach occurs. Both data at rest (stored on servers) and data in transit (being transmitted over networks) should be encrypted using industry-standard encryption algorithms. Key management is critical; secure storage and rotation of encryption keys are necessary to maintain data security.

Consider data anonymization and pseudonymization techniques for protecting sensitive personal data. These methods replace identifying information with pseudonyms or remove identifying information altogether, allowing for analysis without compromising individual privacy. This is particularly relevant when working with client data. Compliance with data privacy regulations necessitates the implementation of such techniques.

Regular data backups are a crucial part of any data management strategy. Backups provide a safeguard against data loss due to hardware failures, software errors, or cyberattacks. Implement a robust backup and recovery strategy, regularly testing the restoration process to ensure its effectiveness. This should include both on-site and off-site backups to protect against physical damage or theft.

The process of data cleaning and preprocessing is often overlooked, but it is critical to the success of any AI-powered investment strategy. Raw data rarely comes in a perfectly usable format; it usually contains errors, inconsistencies, and missing values. Data cleaning involves identifying and correcting these issues, ensuring that the data fed into the AI model is accurate and reliable. This might involve handling missing values through imputation, standardizing data formats, and removing outliers. The quality of the data directly impacts the accuracy of the AI's predictions; inaccurate data invariably leads to inaccurate results. Thorough data validation should be performed at each step of the process.

Developing a comprehensive data governance framework is essential for maintaining data quality and security. This framework should define roles, responsibilities, and processes related to data management, ensuring accountability and consistent application of best practices. Regularly review and update your data governance framework to adapt to changes in technology, regulations, and business needs.

AI model explainability and interpretability are critical considerations in the financial industry. Regulators often require transparency in algorithmic decision-making, particularly in areas such as credit scoring and fraud detection. Choose AI models that offer greater explainability, such as linear regression or decision trees, over "black box" models like deep neural networks, where it's more challenging to understand the reasoning behind the model's predictions. Documentation of the model's training process, features used, and performance metrics is crucial for meeting regulatory requirements and building trust.

Finally, remember that data management and security are not one-time tasks; they are ongoing processes that require continuous monitoring, evaluation, and improvement. Regular security assessments, updates to your software and infrastructure, and employee training are essential to maintaining a secure and compliant data environment. Staying current with industry best practices and emerging threats is crucial for protecting your data and ensuring the long-term success of your AI-powered investment tools. The investment in robust data management and security is not an expense but a critical investment in the overall success and longevity of your AI-driven investment strategy. Neglecting this aspect jeopardizes not only financial returns but also the firm's reputation and legal standing.

Understanding and interpreting the outputs of AI models is a crucial, often overlooked, skill in the realm of AI-powered investing. While AI can process vast datasets and identify patterns invisible to the human eye, its outputs are not infallible pronouncements. Treating AI predictions as gospel can lead to significant losses. The true value lies in understanding the nuances of the output, its inherent limitations, and how to integrate it effectively into your overall investment strategy. This requires a blend of technical understanding and sound financial judgment.

One of the primary challenges lies in the "black box" nature of some AI models. Deep learning algorithms, for example, are notorious for their opacity. While they can achieve impressive predictive accuracy, understanding

why they arrive at a specific prediction can be extremely difficult. This lack of transparency is problematic in the finance industry, where regulatory scrutiny and accountability are paramount. Explainable AI (XAI) techniques are emerging to address this challenge, aiming to make the decision-making process of AI models more transparent and understandable. However, even with XAI, complete interpretability remains a challenge for some complex models.

Therefore, it's crucial to select AI models with a reasonable degree of transparency. Simpler models, such as linear regression or decision trees, offer greater explainability. These models provide a clear picture of the relationships between input variables and the resulting predictions. While their predictive power may not always match that of more complex models, the insights gained from their interpretability are invaluable. This allows for a more thorough understanding of the drivers behind the AI's suggestions, facilitating a better informed investment decision.

Beyond the model's inherent explainability, understanding the context of the AI's output is vital. The AI's predictions are only as good as the data it's trained on. Bias in the training data can lead to skewed or inaccurate predictions. For instance, if the model is trained on historical data from a bull market, it might be overly optimistic in predicting future returns, failing to account for potential market downturns. Similarly, if the data is incomplete or contains errors, the

AI's predictions will reflect these shortcomings. A thorough understanding of the data used to train the model, its limitations, and potential biases is crucial for accurate interpretation.

Furthermore, it's crucial to remember that AI predictions are probabilistic, not deterministic. An AI model might predict a 70% probability of a stock price increase, but this doesn't guarantee a price rise. There's always a degree of uncertainty associated with any prediction, particularly in the volatile world of finance. Understanding this probabilistic nature helps in managing risk and making more informed decisions. Over-reliance on probabilities without considering other market factors can lead to disastrous outcomes.

It is equally important to understand the limitations of the data itself. Financial markets are complex adaptive systems, influenced by a vast range of factors beyond the scope of even the most sophisticated datasets. Unexpected geopolitical events, regulatory changes, and shifts in consumer sentiment can all dramatically impact market performance, and these events are often unpredictable and difficult to incorporate into AI models. Therefore, relying solely on AI predictions without considering these broader contextual factors can be shortsighted and even reckless.

Another significant consideration is the potential for overfitting. Overfitting occurs when an AI model becomes too closely tied to the specific characteristics of the training data, leading to poor performance on new, unseen data. An overfit model might achieve high accuracy on the training data but fail miserably when applied to real-world investment scenarios. This highlights the importance of rigorous testing and validation of the AI model using separate datasets, ensuring its generalizability and robustness.

Moreover, the interpretation of AI outputs shouldn't be a purely quantitative exercise. Qualitative factors, such as management quality, competitive landscape, and regulatory environment, also play a critical role in investment decisions. While AI can provide valuable quantitative insights, it shouldn't replace fundamental analysis and sound financial judgment. AI should be viewed as a powerful tool to augment human expertise, not replace it. Integrating AI insights with traditional investment approaches leads to more robust and informed decision-making.

Let's illustrate these points with a real-world example. Imagine an AI model predicting the performance of a particular stock based on historical price data and various financial indicators. The model might predict a high probability of price appreciation. However, if subsequent analysis reveals that the company is facing a significant lawsuit or has been implicated in unethical practices, the AI's prediction might be severely misleading. This exemplifies the need to consider qualitative factors alongside quantitative data from AI models. A well-rounded investment decision integrates these insights, offering a more comprehensive and robust assessment.

Another common pitfall is relying solely on a single AI model's output. Diversification applies to AI models as much as it does to investment portfolios. Utilizing multiple AI models with different methodologies and data sources can provide a more comprehensive and robust view of the market. The convergence or divergence of predictions from different models provides valuable insights into the reliability and potential limitations of the individual predictions. This approach reduces the risk of making decisions based on the idiosyncrasies of a single model.

Finally, successful utilization of AI in investment decision-making requires a continuous learning approach. Regularly monitoring the performance of the AI models, refining their parameters as needed, and incorporating new data sources are essential for ensuring the continued accuracy and relevance of the AI's predictions. The ever-evolving nature of financial markets demands a dynamic and adaptable approach to AI implementation. Staying abreast of new AI techniques, data sources, and market trends is critical for maintaining a competitive edge. The integration of AI in investment analysis is an ongoing process, not a one-time project. Continuous adaptation and refinement are key to long-term success.

In conclusion, effectively using AI in investments requires more than just plugging data into a model and blindly accepting its output. Success hinges on a deep understanding of the AI model's capabilities and limitations, a critical evaluation of the data used, a balanced approach that integrates both quantitative and qualitative factors, and a continuous learning process that adapts to the dynamism of the financial markets. The synergy between human expertise and AI capabilities, with a focus on responsible interpretation of AI outputs, is the key to unlocking the true potential of AI in investment management. This nuanced understanding transforms AI from a mere tool into a strategic partner in navigating the complexities of the financial world.

The integration of AI into investment strategies is not a static process; it's a dynamic journey requiring constant learning and adaptation. The financial landscape is in perpetual motion, shaped by unpredictable geopolitical events, technological breakthroughs, and shifting economic tides. To remain competitive and effectively leverage AI's potential, a commitment to continuous learning is paramount. This commitment extends beyond simply understanding the current capabilities of AI; it demands a proactive engagement with the evolving field of artificial intelligence and its applications within finance.

One of the most crucial aspects of continuous learning is staying abreast of the latest advancements in AI algorithms and methodologies. The field is rapidly evolving, with new models and techniques emerging at an astonishing pace. Deep learning, once a niche area, has now become a mainstream technology, constantly being refined and improved. Reinforcement learning, a powerful method for training AI agents to make optimal decisions in dynamic environments, is gaining traction in financial applications, offering the potential to optimize portfolio construction and risk management in ways previously unimaginable. Staying informed about these advancements requires a dedicated effort to read research papers, attend industry conferences, and participate in online communities dedicated to AI in finance. This is not merely an intellectual exercise; it is a critical component of maintaining a competitive edge.

Beyond algorithms, the continuous improvement of data sources and their integration into AI models is equally crucial. New data sources, from alternative data providers to social media sentiment analysis, are constantly emerging, offering potentially valuable insights into market trends and investor behavior. The ability to effectively integrate and interpret this data is becoming increasingly important. This requires not only technical proficiency but also a keen understanding of the limitations and potential biases inherent in these data sets. Understanding data quality, handling missing data, and addressing biases are crucial skills for anyone working with AI in finance. The challenge lies not only in accessing these data sources but also in effectively filtering, cleaning, and preprocessing the data to ensure its suitability for AI model training and deployment.

Furthermore, the continuous learning process should encompass a critical evaluation of the performance of existing AI models. Regular monitoring and evaluation are essential to ensure that the models continue to meet the desired performance criteria. This evaluation involves tracking key metrics such as accuracy, precision, recall, and F1-score, adapting thresholds as necessary, and consistently benchmarking against other models and strategies. If the performance degrades over time, this indicates the need for model retraining or even the adoption of a new model altogether. This iterative process of testing, evaluation, and refinement is vital for maintaining the accuracy and relevance of AI-powered investment tools.

Adapting strategies based on the latest advancements and feedback is another critical element. The efficacy of an AI-powered investment strategy is dependent on its ability to adapt to evolving market conditions. This might involve adjusting the model parameters, modifying the data input, incorporating new features, or even switching to a completely different AI model altogether. A rigid adherence to a fixed strategy, regardless of changing market conditions or new insights, can quickly become obsolete and even detrimental. Flexibility and adaptability are paramount for long-term success.

The process of adapting strategies is not solely a technical endeavor; it requires a thorough understanding of market dynamics, regulatory changes, and macroeconomic factors. While AI can provide valuable quantitative insights, it's crucial to integrate this information with qualitative factors and traditional financial analysis to arrive at a well-rounded investment decision. This integration of AI and human expertise is vital to mitigating the risks inherent in relying solely on automated systems.

To facilitate this ongoing learning process, several resources are readily available. Numerous online courses, workshops, and webinars specifically address the application of AI in finance. Platforms like Coursera, edX, and Udacity offer structured learning paths covering various aspects of AI, machine learning, and data science. These courses can provide the foundational knowledge necessary to understand and utilize the latest AI techniques. Furthermore, attending industry conferences and workshops offers the opportunity to network with other professionals in the field, learn about cutting-edge research, and exchange best practices. Active participation in online communities dedicated to AI in finance, through forums and discussions, allows for continuous learning and collaboration with experts from around the globe.

The world of academic research in AI and finance is another invaluable source of information. Regularly reviewing publications from leading journals and research institutions provides access to the latest findings and breakthroughs in the field. While the technical depth of these publications may vary, staying abreast of these advancements can enhance your understanding of the theoretical underpinnings of AI algorithms and their practical applications in financial markets.

Beyond formal education, a commitment to self-learning is essential. Reading industry publications, following thought leaders on social media, and staying updated on news related to AI and finance are crucial for continuous learning. Experimentation is also a powerful tool; testing new algorithms, exploring new data sources, and implementing different models can provide invaluable practical experience and accelerate the learning process. This proactive approach to learning, beyond formal courses and workshops, fosters a deep understanding of the practical aspects of implementing AI in finance.

In summary, building and maintaining a successful AI investment toolkit requires a continuous learning process that adapts to the ever-evolving landscape of AI and finance. This involves staying updated on algorithmic advancements, leveraging new data sources, critically evaluating model performance, and adapting strategies based on feedback and market dynamics. By proactively utilizing the available resources and fostering a commitment to self-learning, you can ensure that your AI investment toolkit remains sharp, relevant, and effective in navigating the complexities of the financial world. The success of AI-powered investing hinges not only on the technical prowess of the algorithms but also on the continuous learning and adaptability of the individuals who employ them. This dynamic engagement is the key to unlocking the full potential of AI in financial decision-making.

The integration of artificial intelligence (AI) into investment strategies, while offering immense potential, introduces a new layer of complexities and risks that require careful consideration and robust mitigation strategies.

Ignoring these risks can lead to significant financial losses, reputational damage, and regulatory repercussions. This section delves into the identification and assessment of key AI-related risks in the investment landscape.

One of the most significant risks stems from algorithmic bias. AI models are trained on historical data, and if this data reflects existing biases—whether conscious or unconscious—the resulting model will likely perpetuate and even amplify these biases. For instance, a model trained on data predominantly representing a specific demographic might produce biased investment recommendations, potentially overlooking opportunities or disproportionately favoring certain asset classes or sectors. This can lead to unfair or discriminatory outcomes and expose firms to legal and ethical challenges. Consider a loan approval algorithm trained on historical data showing a higher default rate among applicants from a particular socioeconomic background. Without careful mitigation, the algorithm might unfairly deny loans to applicants from this group, even if their individual creditworthiness is sound. The consequences could include significant financial harm to individuals and reputational damage to the institution deploying the algorithm. Furthermore, regulators are increasingly scrutinizing algorithmic fairness, and biased models can lead to hefty fines and other penalties. Mitigating this risk requires careful data curation, the selection of diverse and representative datasets, and the implementation of fairness-aware algorithms. Regular audits and testing for bias are also essential.

Model inaccuracy is another critical risk. AI models, even the most sophisticated ones, are not perfect. They can make errors in their predictions and recommendations, leading to potentially significant financial losses. The complexity of financial markets, the presence of "black swan" events, and the inherent uncertainty of future market movements all contribute to the challenges of creating perfectly accurate predictive models. Overreliance on AI models without incorporating human oversight and judgment can exacerbate this risk. Imagine an AI-powered trading algorithm that misinterprets a news report, leading it to make a series of ill-advised trades based on erroneous information. Such scenarios underscore the importance of robust validation procedures, stress testing, and continuous monitoring of model performance. It's also crucial to establish clear parameters for human intervention to override AI recommendations when necessary. Maintaining a skeptical yet constructive attitude towards AI outputs, coupled with a rigorous methodology for backtesting and validation, is crucial in minimizing this risk. Regular reviews of model assumptions and their alignment with evolving market conditions are vital.

Data breaches pose a considerable threat in the context of AI-driven investing. AI models often rely on vast amounts of sensitive data, including personal financial information, market data, and proprietary trading strategies. A data breach can expose this sensitive information to malicious actors, leading to financial losses, identity theft, and reputational harm. The consequences can be particularly severe if the breach exposes trade secrets or proprietary algorithms. The financial industry is a prime target for cyberattacks, and the use of AI only amplifies the potential impact of a successful breach. Implementing robust cybersecurity measures is non-negotiable. This includes encryption, access controls, intrusion detection systems, and regular security audits. Furthermore, the choice of cloud providers and data storage solutions must be meticulously evaluated, selecting providers with strong security credentials and a proven track record in safeguarding sensitive data. Employee training in cybersecurity best practices is equally vital to prevent breaches caused by human error. It's imperative to implement a comprehensive incident response plan to effectively manage and mitigate the consequences of a data breach, including notification procedures and damage control strategies. This proactive approach to cybersecurity will be essential in protecting not just financial assets, but also the reputation of the firm and the trust placed in its AI systems.

Beyond these core risks, there are several other crucial considerations. The "black box" nature of some AI algorithms can make it difficult to understand how they arrive at their conclusions. This lack of transparency can make it challenging to identify and address errors or biases in the model. Explainable

AI (XAI) is emerging as a critical field aimed at making AI algorithms more transparent and interpretable. XAI techniques help uncover the rationale behind AI decisions, facilitating both debugging and risk management. Another emerging concern is the potential for AI systems to be manipulated or gamed. Malicious actors might attempt to manipulate data inputs or exploit vulnerabilities in the algorithms to gain an unfair advantage. Continuous monitoring and security updates are crucial in mitigating this threat. Furthermore, a significant risk is the potential for regulatory changes to impact the use of AI in investing. As AI becomes more pervasive in the financial industry, regulators are developing new rules and guidelines to address the risks associated with its use. Staying abreast of evolving regulations and ensuring compliance is crucial to avoiding fines and penalties.

In conclusion, integrating AI into investment strategies necessitates a proactive and comprehensive approach to risk management. It is not enough to simply acknowledge these risks; a robust framework must be developed to identify, assess, and mitigate them. This framework should include rigorous testing procedures, continuous monitoring of model performance, and a clear process for human oversight. Furthermore, it's crucial to prioritize data security, develop strong cybersecurity protocols, and maintain up-to-date knowledge of relevant regulations. The successful implementation of AI in investing depends not only on the sophistication of the technology but also on the effectiveness of the risk management strategies employed. The continuous learning process discussed previously is essential for staying ahead of emerging risks and adapting risk management strategies to reflect evolving threats and opportunities. This ongoing adaptation, coupled with a culture of transparency and robust security protocols, will be vital in unlocking the transformative potential of AI in the investment world while mitigating the inherent risks.

Mitigating algorithmic bias is paramount in ensuring the fairness, reliability, and ethical use of AI in investment strategies. The potential for AI models to perpetuate and amplify existing societal biases is a significant concern, potentially leading to discriminatory outcomes and substantial reputational damage for financial institutions. This risk is not merely theoretical; real-world examples demonstrate how biased algorithms can lead to unfair loan denials, skewed investment recommendations, and ultimately, financial harm to individuals and communities. Addressing this requires a multifaceted approach encompassing data preprocessing, algorithmic design, and ongoing monitoring and evaluation.

The foundation of mitigating bias lies in the data used to train AI models. Biased data inevitably leads to biased models. Therefore, meticulous attention must be paid to data sourcing, cleaning, and representation. A diverse and representative dataset is crucial. This means actively seeking out data points that represent a broad spectrum of demographics, socioeconomic backgrounds, and market conditions. Over-reliance on easily accessible or historically dominant datasets can inadvertently perpetuate existing biases. For instance, if a model predicting loan defaults is primarily trained on data from affluent neighborhoods, it might incorrectly assume that individuals from less affluent areas are inherently riskier borrowers.

Data preprocessing techniques play a crucial role in mitigating bias. This involves not just cleaning the data – removing duplicates, handling missing values, and correcting errors – but also proactively identifying and addressing potential biases. Techniques such as re-weighting, data augmentation, and adversarial training can help balance the dataset and reduce the impact of skewed representations. Re-weighting involves assigning different weights to data points based on their representation in the dataset, giving more weight to underrepresented groups. Data augmentation involves creating synthetic data points to increase the representation of underrepresented groups, while adversarial training involves training the model to be robust against biased inputs. Each of these techniques requires careful consideration and application to avoid inadvertently introducing new biases or over-correcting existing ones.

Beyond data preparation, the choice of algorithm itself significantly influences the potential for bias. Some algorithms are inherently more susceptible to bias than others. For example, simpler algorithms might lack the capacity to capture complex interactions within the data, leading to oversimplifications that perpetuate bias. More sophisticated models, while potentially more accurate, can be harder to interpret and therefore more challenging to audit for bias. The selection of an appropriate algorithm requires a careful consideration of its inherent properties, its capacity to handle complex datasets, and its interpretability. Focusing on explainable AI (XAI) models, which offer insights into the reasoning behind their predictions, is crucial in detecting and mitigating bias.

Regular auditing and testing for bias are essential components of a comprehensive risk management strategy. This should not be a one-time exercise but rather an ongoing process integrated into the model's lifecycle. A range of techniques can be used to assess fairness, including examining the model's predictions across different demographic groups, assessing the model's fairness metrics such as disparate impact and equal opportunity, and employing statistical tests to detect significant disparities. The choice of specific metrics and testing methods should depend on the specific context and goals of the AI system.

Moreover, incorporating human oversight is crucial in mitigating algorithmic bias. While AI can automate many aspects of investment analysis, human experts should play a vital role in reviewing model outputs, challenging assumptions, and identifying potential biases. This requires professionals with both technical expertise and a strong understanding of ethical considerations and potential biases. Their involvement is not about replacing AI but about augmenting its capabilities and ensuring responsible use.

The financial industry is increasingly subject to regulatory scrutiny regarding the fairness and transparency of AI-driven systems. Regulators are developing guidelines and standards to address algorithmic bias, and adherence to these regulations is not only essential for compliance but also critical for maintaining the trust and confidence of stakeholders. Staying informed about evolving regulatory landscapes and incorporating best practices for algorithmic fairness is a vital part of responsible AI development.

Beyond technical solutions, fostering a culture of responsible AI development within financial institutions is crucial. This involves training employees on the ethical implications of AI, promoting diversity and inclusion in development teams, and establishing clear procedures for identifying and addressing biases. Creating a transparent and collaborative environment where concerns about bias can be raised and addressed without fear of retribution is essential for building trust and promoting responsible innovation.

In conclusion, mitigating algorithmic bias in AI-powered investment tools is a complex challenge demanding a multi-pronged approach. It begins with meticulous data preparation, extends through careful algorithm selection and rigorous testing, and culminates in ongoing monitoring and the integration of human expertise. The financial industry must prioritize transparency, fairness, and ethical considerations throughout the entire AI lifecycle to ensure that AI's transformative potential benefits all stakeholders and does not exacerbate existing inequalities. This commitment to responsible AI development is not merely a matter of ethical responsibility; it's a necessary step for achieving sustainable growth and avoiding the significant financial and reputational risks associated with biased algorithms. The continuous refinement of techniques and the ongoing dialogue among researchers, practitioners, and regulators are vital to ensuring the ethical and effective use of AI in finance. The evolving nature of AI necessitates a constant reassessment of mitigation strategies, reflecting advancements in technology and a deepened understanding of the societal impacts of these powerful tools. Only through this ongoing commitment to responsible innovation can the financial industry fully realize the potential of AI while minimizing its inherent risks.

Managing model uncertainty is crucial for responsible AI implementation in finance. AI models, however sophisticated, are not crystal balls; they provide predictions based on the data they are trained on and the algorithms employed. This inherent uncertainty, often overlooked, can significantly impact investment decisions if not properly accounted for. Failing to acknowledge this uncertainty can lead to overconfidence in AI-driven predictions, resulting in potentially disastrous investment strategies. A robust risk management framework must therefore incorporate methods for quantifying and managing this inherent model uncertainty.

One primary approach to managing model uncertainty is through probabilistic forecasting. Instead of providing single-point predictions, AI models can be adapted to generate probability distributions. This means, instead of stating that a particular stock will reach a certain price, the model would provide a range of possible outcomes along with associated probabilities. For example, rather than predicting a stock price of $150, a probabilistic model might predict a 60% probability of the stock price falling between $140 and $160, a 30% probability of it being between $130 and $140, and a 10% probability of it exceeding $160. This provides investors with a much more nuanced understanding of the potential risks involved.

This approach requires careful consideration of the underlying data and model assumptions. The accuracy of the probability distributions depends heavily on the quality and representativeness of the training data. If the data used to train the model doesn't adequately capture the full range of market conditions, the resulting probability distributions will likely be inaccurate and unreliable. Therefore, rigorous data validation and preprocessing are paramount in ensuring the reliability of probabilistic forecasts. Furthermore, the choice of probabilistic model also plays a crucial role. Different models offer varying degrees of complexity and accuracy, requiring careful consideration based on the specific application and available data.

Beyond probabilistic forecasting, ensemble methods offer a powerful approach to managing model uncertainty. Ensemble methods involve combining predictions from multiple individual models. By aggregating the predictions from various models trained on different subsets of data or using different algorithms, the overall prediction accuracy can be improved and model uncertainty can be reduced. The underlying principle is that the strengths of some models will compensate for the weaknesses of others, reducing the overall risk associated with relying on a single model's prediction. However, it's important to note that the complexity of ensemble methods increases significantly with the number of individual models included. This necessitates careful consideration of computational resources and the potential for diminishing returns as more models are added.

Sensitivity analysis is another crucial technique for managing model uncertainty. This involves systematically varying the input parameters of the AI model to observe the impact on the output predictions. This helps to identify which parameters are most influential in shaping the model's predictions and highlight areas of high uncertainty. For instance, if a model predicting portfolio returns is highly sensitive to changes in interest rates, this highlights a significant source of uncertainty and prompts a deeper investigation into the potential impact of different interest rate scenarios. The results of a sensitivity analysis can inform the development of robust investment strategies that account for potential variations in key input parameters.

Scenario planning, closely related to sensitivity analysis, involves developing a range of potential future scenarios based on different assumptions about key variables. These scenarios can be used to test the robustness of investment strategies under various conditions. For example, an investment strategy might be tested under scenarios including a bull market, a bear market, a period of high inflation, or a geopolitical crisis. This process helps to identify potential vulnerabilities and adjust investment strategies accordingly. Scenario planning complements probabilistic forecasting by providing a more narrative-driven approach to understanding potential outcomes and their implications.

Stress testing, a widely used technique in risk management, takes a more extreme approach. It involves subjecting the AI model and associated investment strategies to extreme or hypothetical events. This might involve testing the model's performance under conditions significantly different from those observed in historical data. For example, a stress test might involve simulating a sudden and sharp drop in market values or an unexpected geopolitical event. This helps to identify potential weaknesses in the investment strategy and areas where greater risk mitigation is required.

In addition to these technical approaches, incorporating human expertise is crucial in managing model uncertainty. AI models can provide valuable insights, but they shouldn't replace the judgment and experience of human investment professionals. Human experts can provide valuable context, challenge model assumptions, and incorporate qualitative factors that are not easily captured by quantitative models. Their ability to interpret model outputs within a broader economic and market context is invaluable in mitigating the risk of overreliance on AI predictions. The combination of AI and human expertise creates a synergistic approach to investment management that leverages the strengths of both.

Transparency and explainability are also crucial aspects of managing model uncertainty. Understanding how an AI model arrives at its predictions is essential in assessing its reliability and identifying potential sources of error. Explainable AI (XAI) techniques help to make the decision-making process of AI models more transparent, allowing investors to understand the rationale behind the model's predictions. This enhanced transparency increases trust in the model's outputs and facilitates the identification of potential biases or limitations. Without transparency, it becomes difficult to assess the reliability of the model's predictions and manage the associated uncertainty effectively.

Furthermore, continuous monitoring and evaluation of the AI model's performance are critical. The accuracy and reliability of AI models can degrade over time as market conditions change or new data becomes available. Regular monitoring allows for the timely identification of model drift or other performance issues, prompting necessary recalibration or retraining of the model. This continuous improvement cycle is vital in maintaining the reliability of AI-driven predictions and effectively managing model uncertainty.

The regulatory landscape is also playing an increasingly significant role in managing model uncertainty. Regulators are increasingly requiring financial institutions to provide detailed explanations of how their AI systems work and assess their associated risks. Compliance with these regulations not only ensures adherence to the law but also fosters greater transparency and accountability in the use of AI in finance. Staying abreast of these evolving regulations is crucial for financial institutions to manage their risk effectively and maintain the trust of their stakeholders.

In conclusion, managing model uncertainty is not a single, isolated task but rather an ongoing and iterative process that requires a holistic approach. It necessitates the use of advanced statistical techniques, such as probabilistic forecasting and ensemble methods, coupled with rigorous testing procedures like sensitivity analysis and stress testing. The integration of human expertise, a commitment to transparency and explainability, and continuous monitoring and evaluation are all crucial components of a robust risk management framework. By embracing these approaches, financial institutions can harness the power of AI while effectively managing the inherent uncertainties of these powerful tools, leading to more informed, responsible, and ultimately more successful investment strategies. The future of finance will undoubtedly involve AI, but its effective and safe implementation hinges on the ability to acknowledge and manage the inherent uncertainties that accompany this transformative technology.

The previous section detailed strategies for managing the inherent uncertainties of AI models in finance. However, the power of AI in finance comes at a cost: the increased risk associated with handling vast quantities of sensitive financial data. This section focuses on the critical issues of data security and privacy in AI-powered investing, exploring best practices and compliance strategies to safeguard sensitive information. The very essence of AI's predictive capabilities rests upon the data it consumes; thus, protecting this data is paramount to the integrity and trustworthiness of the entire system.

The sheer volume of data involved in AI-driven finance is staggering. We're talking about transactional records, personal financial details, market data, and countless other pieces of information, all needing robust protection. A breach could have catastrophic consequences, ranging from financial losses and reputational damage to legal penalties and erosion of customer trust. This is not merely a technical challenge; it is a fundamental ethical and legal responsibility. The implications of a data breach extend far beyond the immediate financial impact, often resulting in long-term damage to a firm's reputation and its relationship with clients. In the highly regulated financial world, non-compliance can lead to hefty fines and potential legal action.

One of the first lines of defense is robust data encryption. This involves transforming data into an unreadable format, rendering it useless to unauthorized individuals. Various encryption methods exist, each with varying levels of security. The choice of encryption method should depend on the sensitivity of the data and the threat landscape. For example, data at rest (stored on servers or databases) might require different encryption techniques than data in transit (being transmitted over a network). Furthermore, regular key rotation—periodically changing the encryption keys—adds another layer of protection against potential breaches. Sophisticated encryption techniques, such as homomorphic encryption, allow computations to be performed on encrypted data without decryption, offering an additional safeguard for sensitive information.

Beyond encryption, access control mechanisms are crucial. This involves implementing strict policies that limit access to sensitive data based on the principle of least privilege. Only authorized personnel should have access to specific data sets, and their access should be carefully monitored and logged. Role-based access control (RBAC) is a common approach, assigning different levels of access based on an individual's role within the organization. Multi-factor authentication (MFA), requiring multiple forms of verification (e.g., password, one-time code, biometric scan) for access, enhances security by adding an extra hurdle for potential intruders. Regular audits of access logs are essential to detect any suspicious activity.

Data loss prevention (DLP) systems play a vital role in preventing sensitive data from leaving the organization's control. These systems monitor data flows, identifying and blocking any attempts to transmit sensitive information outside approved channels. This includes monitoring email traffic, file transfers, and cloud storage access. DLP systems can be configured to identify specific data patterns, keywords, or file types, providing a granular level of control over data movement.

Network security is another critical aspect. Firewalls, intrusion detection systems (IDS), and intrusion prevention systems (IPS) are essential components of a robust security infrastructure. These systems monitor network traffic, identifying and blocking malicious activity. Regular security audits and penetration testing simulate real-world attacks to identify vulnerabilities before they can be exploited. These exercises are crucial for proactively identifying and addressing weaknesses in the system's defenses. Furthermore, keeping software up to date and patched is crucial. Outdated software frequently contains known vulnerabilities, making systems more susceptible to attack.

Cloud security is particularly crucial given the increasing reliance on cloud-based services. When using cloud providers, it's vital to carefully evaluate their security measures and choose providers with strong security track records. Implementing strong access controls and encryption for data stored in the cloud is equally vital. The shared responsibility model, where both the cloud provider and the user are responsible for certain aspects of security, must be carefully managed to ensure that both parties fulfill their obligations.

Compliance with relevant regulations is non-negotiable. Regulations such as GDPR (General Data Protection Regulation) in Europe, CCPA (California Consumer Privacy Act) in California, and other regional and national laws dictate stringent requirements for data handling. These regulations specify how organizations must collect, process, and protect personal data. Non-compliance can result in severe penalties. Organizations must develop comprehensive data privacy policies and procedures to ensure compliance, including data mapping exercises to identify all personal data held by the organization and establishing processes for managing data subject requests (e.g., requests to access, rectify, or delete data).

Employee training is also essential. Employees must be educated on data security best practices, including password security, phishing awareness, and the importance of reporting any suspicious activity. Regular security awareness training can significantly reduce the risk of human error, which is often a major contributor to security breaches. A culture of security awareness must be fostered within the organization, ensuring that all employees understand their role in protecting sensitive data.

Data anonymization and pseudonymization techniques offer a further layer of protection. Anonymization removes identifying information from data, making it impossible to link the data back to individuals. Pseudonymization replaces identifying information with pseudonyms, allowing data analysis while preserving privacy. These techniques are particularly useful when sharing data with third parties or for research purposes.

Finally, incident response planning is crucial. Despite the best security measures, breaches can still occur. Having a well-defined incident response plan in place is essential for minimizing the impact of a breach. This plan should outline clear procedures for detecting, containing, investigating, and recovering from a security incident. Regular testing of the incident response plan ensures that procedures are effective and up-to-date.

In conclusion, ensuring data security and privacy in AI-powered investing requires a multifaceted approach encompassing robust technical measures, strong governance frameworks, employee training, and regulatory compliance. It's not a one-time fix but an ongoing process of adaptation, continuous improvement, and unwavering vigilance. The responsibility for safeguarding sensitive financial data is not simply a technical issue; it is a fundamental ethical obligation and a critical factor in maintaining the integrity and trustworthiness of the financial system itself. The future of AI in finance depends on the ability of financial institutions to navigate this complex landscape, ensuring that innovation is coupled with responsibility and unwavering protection of sensitive data. Failure to do so will inevitably lead to significant legal, financial, and reputational consequences.

Building a robust risk management framework for AI in finance requires a systematic approach that goes beyond traditional risk assessment methodologies. The inherent complexities of AI models, coupled with the unique challenges of handling vast datasets and navigating evolving regulations, demand a multi-layered strategy. This framework should encompass identification, assessment, mitigation, monitoring, and reporting of AI-specific risks.

The first step involves comprehensively identifying potential risks. This isn't simply a matter of listing known risks; it requires a proactive, forward-looking approach that considers the unique characteristics of the AI systems deployed.

For instance, model bias, a common issue in AI, can lead to discriminatory lending practices or inaccurate risk assessments. This bias can stem from skewed training data or inherent flaws in the algorithm's design. Another significant risk is model explainability, or the lack thereof – the difficulty in understanding how an AI model arrives at its conclusions. Opaque models can hinder effective risk management, making it difficult to pinpoint the source of errors or miscalculations. Furthermore, data drift – the change in the characteristics of input data over time – can lead to model degradation and inaccurate predictions. A model trained on historical data might become less effective as market conditions evolve or customer behavior shifts. Finally, adversarial attacks, where malicious actors manipulate input data to compromise the model's output, represent a significant security risk.

The next crucial step is risk assessment. This involves evaluating the likelihood and potential impact of each identified risk. This assessment shouldn't be a subjective judgment but should rely on quantitative and qualitative data whenever possible. For example, the likelihood of a model bias leading to significant financial losses can be estimated using historical data and simulations. Similarly, the impact of a data breach can be quantified by considering factors such as the volume of compromised data, the sensitivity of that data, and the potential legal and reputational consequences. A structured risk assessment process, possibly leveraging a risk matrix, should assign priority levels to different risks, focusing resources on the most critical threats.

Once risks are identified and assessed, the focus shifts to mitigation. This requires implementing strategies to reduce the likelihood and impact of these risks. For model bias, mitigation involves careful data curation, using diverse and representative training datasets, and employing techniques like fairness-aware algorithms. For model explainability, techniques such as LIME (Local Interpretable Model-agnostic Explanations) or SHAP (SHapley Additive exPlanations) can help provide insights into the model's decision-making process. Addressing data drift necessitates continuous model monitoring and retraining using updated data. Regular model performance evaluations, including backtesting on historical data and out-of-sample testing on unseen data, are crucial. To mitigate adversarial attacks, robust input validation and anomaly detection techniques should be implemented.

Effective monitoring is the cornerstone of any robust risk management framework. Continuous monitoring of AI models and the data they use allows for early detection of issues such as model degradation, bias emergence, or security breaches. This continuous monitoring may involve implementing automated alerts and dashboards to provide real-time insights into model performance and data quality. This could include real-time monitoring of key performance indicators (KPIs), such as accuracy, precision, recall, and F1-score, alongside data quality metrics. This allows for timely intervention to address any emerging risks. The frequency of monitoring should vary based on the criticality of the model and the dynamic nature of its input data. Higher-frequency monitoring should be implemented for models critical to core business processes and those dealing with sensitive data.

The final element is reporting. Regular reports should summarize the identified risks, the assessment of their likelihood and impact, the implemented mitigation strategies, and the effectiveness of monitoring efforts. These reports provide transparency and accountability and are crucial for improving risk management processes over time. Regular reporting provides stakeholders with a clear understanding of the overall risk profile associated with AI deployment. The reports should be easily understood and should also encompass compliance aspects, ensuring adherence to all relevant regulations. The information presented in these reports must be timely, accurate, and readily accessible to all relevant parties. Senior management, compliance officers, and external auditors should all have access to this critical information.

This comprehensive risk management framework is not static; it requires continuous adaptation. As AI technology evolves and the financial landscape shifts, so too must the risk management strategies. Regular review and updates of the framework are necessary to ensure it remains effective. Furthermore, fostering a culture of risk awareness throughout the organization is critical. Training programs should educate employees about the various risks associated with AI and their roles in mitigating those risks. The framework's success hinges on the commitment and collaboration of all stakeholders, from data scientists and engineers to senior management and compliance officers. This collaborative approach, coupled with the continuous adaptation and improvement mentioned earlier, is critical in navigating the complexities of AI risk management in the dynamic world of finance. The ultimate goal is to harness the power of AI while simultaneously safeguarding against its potential pitfalls, thereby building a sustainable and responsible AI-powered financial future.

Algorithmic bias represents a significant ethical challenge in AI-powered investing. While AI promises to enhance efficiency and potentially democratize access to financial markets, the reality is that biases embedded within algorithms can exacerbate existing inequalities and create new forms of discrimination. These biases, often unintentional, can stem from various sources, ultimately leading to unfair and potentially harmful outcomes for investors. Understanding the nature and impact of algorithmic bias is crucial for building ethical and responsible AI-driven investment systems.

One primary source of bias lies in the training data used to develop AI models. If the data used to train an algorithm reflects existing societal biases, the algorithm will inevitably learn and perpetuate those biases. For example, historical lending data may reflect discriminatory practices of the past, leading to an AI model that unfairly denies loans to certain demographic groups, even if those groups now present a lower risk profile. This isn't a hypothetical concern; numerous studies have demonstrated that AI systems trained on biased data have disproportionately impacted minority communities in areas such as loan applications, insurance underwriting, and even hiring processes within financial institutions. The lack of diversity in the data sets used to train these models further compounds this problem.

Another critical aspect is the design of the algorithms themselves. Even with unbiased data, poorly designed algorithms can introduce biases. For instance, a simple algorithm that prioritizes certain factors over others, without considering their potential for discriminatory impact, can inadvertently lead to unfair outcomes. Consider a wealth management algorithm that solely focuses on maximizing returns, neglecting the risk tolerance and financial goals of diverse investor profiles. Such an algorithm might favor high-risk, high-return investments, potentially harming investors with lower risk tolerance or limited financial resources, disproportionately impacting lower-income individuals. The "black box" nature of some AI models further complicates this issue; understanding how these models arrive at their decisions is often challenging, making it difficult to identify and mitigate biases.

The consequences of algorithmic bias in AI investing are far-reaching. Biased algorithms can lead to inaccurate risk assessments, resulting in financial losses for certain groups of investors. They can perpetuate systemic inequalities by denying access to investment opportunities, limiting financial growth, and widening the wealth gap. This can have significant societal ramifications, contributing to social and economic instability. Imagine a robo-advisor using a biased algorithm to allocate investment portfolios. If this algorithm consistently favors certain asset classes or investment strategies that benefit higher-income individuals, it could exacerbate the existing wealth disparity, creating a system where the rich get richer and the poor struggle to accumulate wealth.

Addressing algorithmic bias requires a multifaceted approach. First and foremost, meticulous attention must be paid to data quality and diversity. This involves actively seeking out and incorporating data that represents the full spectrum of the population, ensuring that no group is underrepresented or misrepresented. Data cleaning and preprocessing techniques can help identify and correct biases present in existing datasets. However, simply removing biased data points may not be sufficient; it's crucial to understand the root causes of the bias to prevent its recurrence.

Furthermore, the development of fairness-aware algorithms is crucial. These algorithms are designed with fairness constraints, ensuring that they do not discriminate against any particular group. Various techniques are available for achieving fairness, each with its own advantages and limitations. For instance, some approaches focus on equalizing outcomes across different demographic groups, while others aim to ensure equal opportunity. The choice of fairness metric depends on the specific context and the desired level of fairness. This requires ongoing research and development to create sophisticated algorithms capable of handling complex datasets and avoiding unintended bias.

Explainability is another vital component. Understanding how an AI model arrives at its decisions is critical for identifying and mitigating bias. Techniques such as LIME and SHAP, as mentioned earlier, can help shed light on the internal workings of complex models, allowing developers and regulators to detect and address any unfairness. This increased transparency also enhances trust in AI-powered investment systems, a crucial aspect for widespread adoption. However, complete explainability is not always feasible, particularly with highly complex models. Therefore, a balance must be struck between model complexity and the need for explainability.

Regulatory oversight plays a critical role in ensuring the ethical use of AI in investing. Regulations can mandate transparency, accountability, and fairness in algorithmic decision-making. This may involve requiring firms to conduct regular audits of their AI systems, to ensure that they are not perpetuating biases. These regulations could also specify requirements for data transparency, algorithmic explainability, and impact assessments, ensuring fairness and preventing discrimination. Furthermore, these regulations should encourage the development and adoption of best practices for building and deploying ethical AI systems. The development of clear guidelines and standards for the responsible use of AI in finance is crucial for fostering trust and promoting ethical innovation.

Finally, a culture of ethical awareness within financial institutions is essential. This requires training programs for data scientists, engineers, and other professionals involved in developing and deploying AI systems. These programs should educate individuals about the potential risks of algorithmic bias, and equip them with the knowledge and skills necessary to mitigate these risks. Furthermore, establishing ethical review boards within organizations can provide an independent assessment of the ethical implications of AI systems before their deployment. Continuous monitoring and evaluation of AI systems post-deployment are also essential to ensure that they remain fair and unbiased over time. The development of a culture of responsible AI is a continuous process requiring ongoing commitment and adaptation.

In conclusion, algorithmic bias presents a significant ethical challenge in AI-powered investing. Addressing this challenge demands a multi-pronged approach involving careful data curation, the development of fairness-aware algorithms, increased transparency and explainability, robust regulatory oversight, and a strong ethical culture within financial institutions. By actively addressing these issues, the financial industry can harness the transformative power of AI while ensuring fairness, equity, and responsible innovation for all investors. The future of AI in finance depends on navigating these ethical complexities responsibly, ensuring that the benefits of this technology are shared equitably, and that its potential harms are mitigated effectively. Only through a collaborative effort involving researchers, developers, regulators, and the broader financial community can we build a truly ethical and inclusive future for AI-driven investing. The stakes are high, impacting not only individual investors but also the stability and fairness of the entire financial system. Ignoring the ethical considerations surrounding algorithmic bias is simply not an option; it's a fundamental requirement for building a sustainable and trustworthy AI-powered financial future.

The ethical considerations surrounding AI in investing extend beyond the issue of algorithmic bias; they delve into the critical need for transparency and explainability in the decision-making processes of these sophisticated systems. While AI promises enhanced efficiency and potentially superior returns, the "black box" nature of many algorithms raises serious concerns about accountability and trust. Understanding *how* an AI system arrives at its investment recommendations is not just a matter of intellectual curiosity; it's a fundamental requirement for responsible deployment and for building public confidence in this rapidly evolving technology. Without transparency and explainability, the potential for unintended consequences, including discriminatory outcomes, remains substantial.

The lack of transparency in some AI systems stems from their inherent complexity. Many AI models, particularly deep learning models, are characterized by a vast number of interconnected parameters and layers. These intricate architectures can make it exceedingly difficult to trace the chain of reasoning that leads to a specific investment decision. This opacity can hinder our ability to identify and correct errors, biases, or flaws in the system's logic. It can also make it challenging to understand the model's vulnerabilities to manipulation or adversarial attacks. Such vulnerabilities pose significant risks, particularly in the high-stakes world of financial markets.

Several techniques are being developed to improve the transparency and explainability of AI models used in investing. These techniques fall broadly into two categories: model-agnostic methods and model-specific methods. Model-agnostic methods can be applied to any type of AI model, regardless of its underlying architecture. One prominent example is Local Interpretable Model-agnostic Explanations (LIME). LIME works by approximating the behavior of a complex model locally around a specific input (e.g., a particular set of financial data). It creates a simpler, more interpretable model that mimics the complex model's behavior within this local region, providing insights into the factors that contributed to the complex model's prediction.

Another popular model-agnostic technique is SHapley Additive exPlanations (SHAP). SHAP uses game theory to assign importance values to each input feature, quantifying the contribution of each feature to the model's overall prediction. By analyzing these importance scores, we can gain a better understanding of the factors driving the AI's investment decisions. Both LIME and SHAP offer valuable tools for investigating the "why" behind an AI-driven investment recommendation, allowing us to identify potential biases or flawed reasoning.

Model-specific methods, on the other hand, are designed to enhance the transparency of particular types of AI models. For instance, in simpler linear models, the weights assigned to each input feature directly indicate the importance of that feature in the model's predictions. This offers a straightforward way to understand the model's logic. However, even with simpler models, a comprehensive understanding requires careful consideration of the data used to train the model. If the training data itself reflects existing biases, the model's interpretations, even if seemingly transparent, might still perpetuate those biases.

Beyond specific techniques like LIME and SHAP, the development of more inherently interpretable models is a crucial area of ongoing research. This involves designing AI architectures that are simpler and easier to understand, without sacrificing their predictive accuracy. This effort seeks to build models that are not only effective but also transparent and readily explainable. For example, decision trees, while perhaps less powerful than deep learning models in some scenarios, offer a highly interpretable representation of the decision-making process. Their structure allows us to clearly trace the path from input to output, identifying the key factors that led to a particular investment recommendation.

The pursuit of transparency and explainability is not merely an academic exercise; it has profound implications for regulatory compliance and investor protection. Financial regulators are increasingly demanding greater transparency from firms using AI in their investment processes. This demand reflects the need to mitigate risks associated with opaque algorithms, ensuring fairness, accountability, and preventing manipulation. Regulations might require firms to provide detailed explanations of how their AI systems make investment decisions, enabling independent audits and assessments of their fairness and accuracy.

Furthermore, enhanced transparency and explainability can foster greater trust among investors. Investors are more likely to trust and adopt AI-powered investment tools if they understand how these systems work and the reasoning behind their recommendations. This trust is essential for the widespread adoption of AI in the financial industry, benefiting both individual investors and the financial system as a whole. A lack of transparency can, conversely, lead to skepticism and reluctance to adopt potentially beneficial AI-driven investment strategies.

However, the drive for transparency and explainability should not come at the cost of model performance. Striking the right balance between interpretability and accuracy is a challenge that requires ongoing research and innovation. The ideal scenario involves developing AI models that are both highly accurate and sufficiently transparent to allow for a thorough understanding of their decision-making processes. The pursuit of this ideal necessitates a collaborative effort among researchers, developers, regulators, and investors.

The development of standards and best practices for transparency and explainability in AI-driven investing is crucial. These standards should be specific to the context of the financial industry, considering the unique requirements for accountability, fairness, and investor protection. These standards should also encourage innovation in AI techniques, ensuring that the pursuit of transparency does not stifle the development of powerful and useful AI models. The ongoing dialogue and collaboration among stakeholders are essential to defining these standards and ensuring their effective implementation.

Finally, it is vital to recognize that complete transparency and explainability might not always be feasible, especially with the most complex AI models. In such cases, a layered approach may be necessary, providing different levels of explanation tailored to the audience and the specific context. For example, a high-level summary of the model's reasoning might be sufficient for a non-technical investor, while more detailed technical explanations could be provided to regulators or expert auditors. The key lies in providing sufficient transparency to ensure accountability and trust, while also acknowledging the inherent limitations of complete explainability in some sophisticated AI systems. The ongoing development of better methodologies for evaluating and communicating model uncertainties is paramount in this pursuit. This includes clear articulation of the model's limitations and the potential for errors in its predictions.

The preceding discussion highlighted the crucial need for transparency and explainability in AI-driven investment strategies. However, even with the most transparent algorithm, the ethical landscape of AI investing remains incomplete without a thorough examination of data privacy and security. The very foundation of AI's power in finance rests on the vast quantities of data it processes – data that often includes highly sensitive personal and financial information belonging to clients. Protecting this data is not just a matter of good business practice; it's a moral imperative and, increasingly, a legal requirement.

The sheer volume and variety of data used in AI-powered investing present significant challenges to data security. These systems often draw on diverse sources, including transactional data, market data, news feeds, social media sentiment, and even alternative data such as satellite imagery or geolocation data. Each data source presents its own unique security vulnerabilities. For instance, transactional data, containing details of individual investments and financial holdings, is a prime target for malicious actors seeking to exploit personal information for financial gain or identity theft. Breaches of such data can lead to severe financial losses for individuals and damage to the reputation of the financial institution involved.

The complexity of AI systems themselves also introduces security risks. The intricate web of algorithms, models, and interconnected databases creates a vast attack surface, offering numerous potential entry points for cybercriminals. Moreover, the use of cloud-based computing, while offering scalability and cost-effectiveness, introduces further security concerns, particularly related to data access and control. Ensuring the integrity and confidentiality of data across diverse cloud environments requires robust security measures and careful management of access privileges.

Regulations governing data privacy and security are evolving rapidly, reflecting the increasing awareness of the risks associated with the use of personal data in AI systems. The General Data Protection Regulation (GDPR) in Europe, for example, sets a high bar for data protection, requiring explicit consent from individuals for the processing of their personal data and providing them with significant control over their data. Similar regulations are emerging globally, including the California Consumer Privacy Act (CCPA) in the United States, aiming to provide consumers with more transparency and control over their personal information.

Compliance with these regulations is not simply a matter of ticking boxes; it requires a holistic approach to data governance, integrating data privacy and security considerations into every stage of the AI investment lifecycle. This includes meticulous data collection and processing procedures, robust data encryption and access control mechanisms, rigorous security audits, and comprehensive incident response plans. Failing to meet these standards can lead to significant financial penalties, reputational damage, and loss of client trust.

Beyond legal compliance, ethical considerations demand a proactive and responsible approach to data privacy and security. Financial institutions deploying AI-powered investment tools have a moral obligation to protect the sensitive data entrusted to them by their clients. This obligation transcends legal requirements, encompassing a commitment to transparency, accountability, and the responsible use of personal data. This includes being upfront with clients about how their data is used, providing clear explanations of the security measures in place, and promptly disclosing any data breaches or security incidents.

The development of robust security protocols is crucial, incorporating best practices such as multi-factor authentication, encryption at rest and in transit, regular security assessments, and penetration testing. AI itself can play a valuable role in enhancing data security, with machine learning algorithms used to detect and prevent fraudulent activities, identify anomalies in data patterns that may indicate security breaches, and automate security monitoring tasks. However, the use of AI in security also presents its own challenges, including the potential for bias in AI-driven security systems and the need for careful oversight of their deployment.

Beyond technical measures, a strong security culture within the organization is paramount. This involves regular training for employees on data privacy and security best practices, promoting a culture of awareness and vigilance, and establishing clear procedures for handling sensitive data. Moreover, establishing a clear chain of accountability for data security is essential, ensuring that individuals are responsible for the data they handle and that mechanisms are in place to investigate and address security incidents promptly.

The ethical considerations extend to the use of data from third-party providers. Financial institutions must carefully vet their data providers, ensuring they adhere to robust security standards and privacy regulations. This includes scrutinizing their data handling procedures, security protocols, and data governance frameworks. Transparency about the sources of data used in AI-powered investment strategies is also crucial, allowing clients to understand the potential risks and implications associated with the use of their data.

The use of anonymization and pseudonymization techniques can help to protect client privacy, replacing identifying information with pseudonyms or removing identifying information entirely while preserving the utility of the data for AI training and analysis. However, it's crucial to acknowledge that these techniques are not foolproof and can introduce challenges in terms of data utility and model accuracy. A careful balance must be struck between privacy protection and the effectiveness of the AI system.

Furthermore, the development of differential privacy techniques is an active area of research, offering a means of protecting individual privacy while still enabling statistical analysis of large datasets. Differential privacy adds noise to data in a way that preserves aggregate statistics while preventing the identification of individual data points. This technique holds significant promise for improving data privacy in AI-powered investing, but its implementation requires careful consideration of the trade-off between privacy and utility.

In conclusion, the ethical use of AI in investing necessitates a robust and multifaceted approach to data privacy and security. This requires not only compliance with existing regulations but also a proactive commitment to ethical data handling practices. This encompasses transparent data governance, robust security measures, and a strong organizational security culture. By prioritizing data privacy and security, financial institutions can build trust with their clients, mitigate risks, and ensure the responsible development and deployment of AI-powered investment strategies. The ongoing evolution of data privacy regulations and the development of new privacy-enhancing technologies will continue to shape the ethical landscape of AI investing, requiring constant vigilance and adaptation from financial institutions.

Building upon the critical discussion of data privacy and security, the responsible development of AI in finance requires a holistic and proactive approach that extends beyond mere compliance. It demands a deep-seated commitment to ethical principles woven into every stage of the AI lifecycle, from initial conception and data acquisition to model deployment and ongoing monitoring. This commitment translates into tangible best practices that shape the development and implementation of AI-powered investment strategies.

One of the most fundamental aspects of responsible AI development is the establishment of clear ethical guidelines and principles. These guidelines should articulate the values and objectives that will guide the development process, ensuring alignment with the broader ethical framework of the financial institution and the industry as a whole. These principles should be readily accessible to all stakeholders, including developers, data scientists, and investment professionals, fostering a shared understanding of ethical responsibilities. The principles should explicitly address issues such as fairness, transparency, accountability, and privacy, providing a clear framework for decision-making throughout the development process. Regular reviews and updates of these ethical guidelines are vital to keep pace with evolving technological capabilities and societal expectations. This continuous adaptation ensures that the AI system remains aligned with the latest ethical standards and best practices.

A critical element of ethical AI development is the careful selection and pre-processing of training data. The quality, representativeness, and bias of the data directly impact the fairness and accuracy of the resulting AI model. Biases present in the training data can perpetuate and amplify existing inequalities, leading to discriminatory outcomes in investment decisions. For example, an AI model trained on historical data that underrepresents certain demographic groups may perpetuate existing biases in investment allocations, potentially disadvantaging these groups. Therefore, rigorous data auditing and bias mitigation techniques are essential. This involves identifying and addressing potential sources of bias in the data, such as skewed sampling or historical biases embedded in financial data. Techniques like re-sampling, data augmentation, and algorithmic fairness constraints can be employed to mitigate bias during model training. Regular audits of the data and the model's performance should be conducted to ensure ongoing fairness and accuracy. Transparency regarding the data sources and the processes used to mitigate bias is crucial for building trust and accountability.

The design and architecture of the AI system itself must incorporate ethical considerations. Explainability and interpretability are paramount, allowing for a clear understanding of how the AI system arrives at its investment recommendations. "Black box" AI models, where the decision-making process is opaque, raise significant ethical concerns, hindering accountability and making it difficult to identify and correct biases or errors. Therefore, efforts should be made to select models that offer greater transparency, such as linear models or decision trees, or to employ techniques like SHAP (SHapley Additive exPlanations) values to explain the contribution of individual features to the model's predictions. This level of transparency enables scrutiny by both internal and external stakeholders, enhancing accountability and enabling the identification of potential biases or inaccuracies.

Furthermore, rigorous testing and validation are crucial to ensure the reliability and robustness of the AI system. Before deployment, the AI model should undergo comprehensive testing using diverse datasets to assess its performance across different scenarios and market conditions. Stress testing the model under extreme conditions can reveal potential vulnerabilities and limitations. This meticulous testing process helps to identify and mitigate potential risks associated with model deployment, safeguarding against unexpected behaviors or erroneous investment recommendations. Continuous monitoring and evaluation after deployment are equally vital, allowing for the detection of any unforeseen biases or performance degradation. This ongoing monitoring enables timely intervention and adjustments, ensuring the AI system maintains its ethical performance and accuracy over time.

The development team's composition plays a significant role in promoting responsible AI development. Diversity within the development team is crucial for ensuring that diverse perspectives are considered throughout the development process. A team composed of individuals with varied backgrounds, experiences, and expertise can better anticipate and address potential biases and ethical considerations. This ensures that the AI system is not merely a reflection of the biases of a homogenous group but rather a product of a more inclusive and representative perspective. Furthermore, regular ethical reviews and consultations with external experts, such as ethicists and legal professionals, can provide valuable insights and perspectives, helping to identify potential ethical risks and ensuring compliance with relevant regulations and best practices.

Beyond the technical aspects, responsible AI development in finance necessitates strong governance and accountability structures. Clear lines of responsibility must be established, outlining who is accountable for the ethical implications of the AI system's decisions and actions. This includes establishing processes for addressing potential ethical dilemmas and resolving conflicts of interest. Regular audits and independent reviews should be conducted to assess compliance with ethical guidelines and regulations. This rigorous oversight ensures that the AI system adheres to ethical standards and that any deviations are promptly addressed. Furthermore, mechanisms for transparent reporting and disclosure are crucial, allowing stakeholders to understand the AI system's capabilities, limitations, and potential impacts. This transparency fosters trust and accountability, crucial for building confidence in the responsible use of AI in the financial industry.

Finally, ongoing education and training for all stakeholders involved in the development and deployment of AI systems are essential. This includes training on ethical considerations, data privacy, bias detection, and responsible AI practices. By fostering a culture of ethical awareness and responsibility within the organization, financial institutions can promote the responsible development and use of AI-powered investment strategies, contributing to a more fair, equitable, and trustworthy financial system. The development and implementation of these measures ensure the ethical and responsible use of AI in finance, building trust, mitigating risks, and fostering a more equitable and transparent financial system. The continuous evolution of technology necessitates an equally evolving commitment to ethical considerations, requiring constant vigilance and proactive adaptation from all stakeholders. Only through a concerted and ongoing commitment to these principles can the promise of AI in finance be realized responsibly and ethically.

The integration of artificial intelligence (AI) into the financial sector holds transformative potential, promising to reshape how we manage money, invest capital, and access financial services. However, this technological leap presents a complex tapestry of social implications, weaving together potential benefits and significant risks that require careful consideration. A failure to address these potential downsides could exacerbate existing inequalities and create new societal challenges.

One of the most compelling potential benefits of AI in finance is increased accessibility to financial services. Traditional banking and investment often have high barriers to entry, leaving many underserved populations—including those in remote areas, low-income individuals, and small businesses—with limited access to crucial financial tools. AI-powered platforms can break down these barriers by automating processes, reducing costs, and creating more personalized and user-friendly interfaces. For example, AI-driven micro-lending platforms can assess creditworthiness more efficiently and accurately than traditional methods, providing access to capital for entrepreneurs in developing countries who might otherwise be excluded from the formal financial system. Similarly, robo-advisors can offer affordable investment advice to individuals who previously couldn't afford the services of human financial advisors, potentially democratizing wealth management and increasing financial literacy across broader segments of society. These examples highlight the capacity of AI to foster financial inclusion and empower individuals who were previously marginalized.

Beyond accessibility, AI can significantly improve the efficiency and transparency of financial markets. AI algorithms can process vast amounts of data to identify market trends, assess risk, and detect fraud far more quickly and accurately than humans can. This enhanced efficiency can translate into lower transaction costs, faster settlement times, and reduced risk for investors. For instance, AI-powered fraud detection systems are continuously evolving, utilizing machine learning to identify and prevent fraudulent transactions in real-time, protecting consumers and institutions from significant financial losses. Moreover, the increased transparency offered by AI-driven systems can enhance market stability and reduce the potential for systemic risks. However, this efficiency boost must be carefully managed to avoid exacerbating existing inequalities. If access to these advanced AI systems is concentrated in the hands of a few large financial institutions, it could further consolidate market power and hinder competition, potentially leading to less favorable outcomes for smaller players and consumers.

The use of AI in algorithmic trading presents both opportunities and risks. AI-powered trading algorithms can execute trades at speeds and frequencies impossible for human traders, enabling the identification and exploitation of fleeting market opportunities. However, this high-frequency trading can contribute to market volatility and potentially increase the likelihood of flash crashes—rapid, unexpected drops in market prices that can cause significant losses for investors. Moreover, the opacity of these algorithms raises concerns about market manipulation and the potential for unfair advantages for those who control them. The lack of transparency in algorithmic trading can make it difficult to understand the drivers of market movements, hindering regulatory oversight and increasing systemic risk. Regulations and oversight mechanisms need to evolve to address these complexities and mitigate the potential negative consequences of AI-driven algorithmic trading, including the potential for market manipulation and the amplification of volatility.

Furthermore, the ethical implications of AI in finance extend to the potential for bias and discrimination. AI models are trained on historical data, and if that data reflects existing societal biases, the AI system will likely perpetuate and even amplify those biases in its decisions. This can lead to discriminatory outcomes in areas like lending, insurance, and investment management, potentially disadvantaging certain demographic groups. For instance, an AI-powered loan application system trained on data that shows historical bias against certain racial or ethnic groups might unfairly deny loans to applicants from those groups, even if they are otherwise creditworthy. Addressing this requires careful consideration of data selection, model design, and ongoing monitoring to identify and mitigate bias. This includes employing techniques like fairness-aware algorithms and regularly auditing models for discriminatory outcomes. Transparency in the data used and algorithms employed is crucial for accountability and for building public trust.

The social impact of AI in finance also encompasses the potential displacement of human workers. As AI systems become more sophisticated, they can automate many tasks currently performed by human financial professionals. This could lead to job losses in various sectors, from customer service and back-office operations to investment analysis and portfolio management. While this concern is valid, it also necessitates proactive strategies to mitigate its consequences. This includes investing in retraining and upskilling programs to equip workers with the skills needed for emerging roles in the AI-driven finance industry. Additionally, policies that promote a just transition to a more automated workforce, including social safety nets and support for displaced workers, are critical to ensure that the benefits of AI are shared broadly and that those negatively impacted are supported. Ignoring this aspect could lead to significant social unrest and economic inequality.

Finally, the concentration of power and control within the financial industry is a serious concern that AI could exacerbate. The development and deployment of advanced AI systems requires substantial financial resources and technical expertise, leading to a potential concentration of power in the hands of large, well-resourced financial institutions. This concentration of power could further marginalize smaller players and limit competition, hindering innovation and potentially leading to less efficient and less equitable outcomes for consumers. Addressing this requires proactive regulatory policies that promote competition, ensure access to AI technologies for smaller firms, and prevent monopolies from forming. Furthermore, promoting open-source development and collaboration in the field of AI finance can contribute to a more inclusive and less centralized ecosystem.

In conclusion, the social impact of AI in finance is a complex and multifaceted issue. While AI has the potential to bring about significant improvements in terms of accessibility, efficiency, and transparency, it also poses risks related to bias, job displacement, and concentration of power. Responsible innovation and proactive mitigation strategies are crucial to ensure that the benefits of AI are broadly shared and that its potential downsides are minimized. This requires a collaborative effort involving governments, financial institutions, researchers, and civil society organizations to establish ethical guidelines, regulatory frameworks, and societal support mechanisms that can guide the development and deployment of AI in finance in a way that benefits all members of society. Without these safeguards, the potential for AI to exacerbate existing inequalities and create new societal challenges remains significant. A thoughtful, proactive approach is paramount to ensuring that the integration of AI into the financial system creates a more equitable, inclusive, and prosperous future for all.

The rapid advancement and integration of artificial intelligence (AI) into the financial sector necessitate a robust and adaptable regulatory framework. The current legal and regulatory landscape, however, is still evolving to address the unique challenges posed by AI's transformative potential. Existing regulations, often designed for traditional financial practices, often struggle to keep pace with the dynamic nature of AI-driven systems. This necessitates a proactive and nuanced approach to regulation, balancing the need to foster innovation with the imperative to mitigate risks and protect consumers.

One of the key challenges lies in the inherent complexity of AI algorithms. Unlike traditional financial models, AI systems, particularly those based on machine learning, are often "black boxes," making it difficult to understand their decision-making processes. This lack of transparency poses significant challenges for regulators seeking to ensure fairness, accountability, and transparency in financial decisions. Regulators are grappling with how to effectively oversee and audit these complex systems, demanding new methodologies and tools for evaluating the performance, fairness, and robustness of AI algorithms employed in finance. This includes developing methods for assessing algorithmic bias and ensuring that these systems do not discriminate against specific groups or perpetuate existing inequalities.

Existing regulations, such as those governing data privacy and consumer protection, play a crucial role in shaping the AI landscape in finance. Regulations like the General Data Protection Regulation (GDPR) in Europe and the California Consumer Privacy Act (CCPA) in the United States establish important standards for data collection, processing, and security. These regulations are crucial in the context of AI, as AI systems rely heavily on vast amounts of data for training and operation. Ensuring compliance with data privacy regulations is critical not only to protect consumer rights but also to maintain trust in the integrity and fairness of AI-driven financial services. The growing emphasis on data governance and responsible data handling presents significant challenges for financial institutions adopting AI, necessitating investments in robust data management systems and compliance protocols. Furthermore, international harmonization of data privacy regulations is becoming increasingly important in a globalized financial system, reducing compliance complexities for international firms.

The regulatory landscape also addresses issues of algorithmic bias and fairness. AI models trained on biased data can perpetuate and even amplify existing societal inequalities, leading to discriminatory outcomes in lending, insurance, and investment management. Regulators are actively exploring ways to mitigate this risk, considering various approaches such as mandating fairness audits of AI systems, promoting the development and use of bias-mitigation techniques, and establishing clear standards for algorithmic transparency and explainability. This includes encouraging the development of "explainable AI" (XAI) systems, which provide insights into how AI systems arrive at their decisions. Achieving this requires a multi-faceted approach, involving collaboration between regulators, AI developers, and researchers, to develop standardized metrics and frameworks for assessing algorithmic fairness and mitigating potential biases.

Another crucial area of regulatory focus is the management of model risk. AI models, by their very nature, are subject to inherent uncertainty and potential errors. Financial institutions using AI systems need robust risk management frameworks to address these potential risks. This includes comprehensive model validation processes, ongoing monitoring of model performance, and contingency plans to manage situations where AI systems fail or produce inaccurate results. Regulators are paying close attention to how financial institutions assess and manage model risk, requiring stringent oversight and reporting mechanisms to ensure the safety and stability of the financial system. This includes developing standards for data quality, model validation, and stress testing to ensure resilience against unexpected events. The integration of AI into financial systems will undoubtedly require a continuous review and updating of existing regulatory frameworks to address the inherent uncertainties and risks associated with complex AI systems.

Furthermore, the use of AI in high-frequency trading presents unique regulatory challenges. The speed and complexity of AI-powered algorithmic trading can increase market volatility and potentially contribute to flash crashes. Regulators are exploring ways to manage this risk, such as implementing transaction taxes, circuit breakers, and enhanced market surveillance mechanisms. The increased computational power and speed of AI-driven trading necessitate a comprehensive evaluation of existing regulatory frameworks, potentially necessitating adjustments to trading protocols and oversight mechanisms to ensure market stability and prevent market manipulation. The opacity of some AI algorithms also raises concerns about the potential for unfair advantages and the difficulty of detecting and preventing market manipulation. Transparency and auditability of algorithmic trading strategies will likely become increasingly important aspects of future regulations.

Beyond specific algorithms, the broader impact of AI on the financial industry's workforce requires attention. The automation potential of AI could lead to job displacement in various roles within the financial sector. This calls for proactive measures, including investment in reskilling and upskilling initiatives to prepare workers for new roles in an AI-driven environment. A forward-looking regulatory approach would encompass strategies for supporting workers impacted by technological advancements, mitigating potential social and economic disruption caused by job displacement. This could involve government-supported training programs, social safety nets, and policies that facilitate the transition of workers to new sectors or roles. The ethical and social implications of AI-driven job displacement demand a comprehensive policy response aimed at minimizing negative impacts and maximizing positive societal outcomes.

The regulatory response to AI in finance also needs to address the potential concentration of power. The development and deployment of advanced AI systems require substantial resources and expertise, potentially creating a significant advantage for large financial institutions. Regulators are concerned about the potential for this to stifle competition and innovation, potentially leading to less efficient and less equitable outcomes for consumers and smaller financial institutions. Addressing this concern might involve implementing measures to promote competition, ensuring access to AI technologies for smaller firms, and preventing the formation of monopolies. Policies promoting open-source development and collaboration in the AI finance space can foster a more inclusive and less centralized ecosystem. A proactive regulatory environment needs to consider the implications of AI on market structure and competition, ensuring that the benefits of AI are broadly shared and not concentrated in the hands of a few dominant players.

In conclusion, the regulatory landscape governing AI in finance is a complex and rapidly evolving area. Existing regulations provide a foundational framework, but substantial adaptation is necessary to effectively address the unique challenges and opportunities presented by AI. This requires a collaborative effort between regulators, financial institutions, AI developers, and researchers to develop a balanced and effective regulatory framework that fosters innovation while mitigating risks and protecting consumers. A proactive, adaptive, and globally harmonized approach is crucial to ensure a safe, fair, and equitable future for AI in the financial sector, maximizing its benefits while minimizing its potential downsides. The ongoing dialogue and collaboration between stakeholders are paramount to establishing a robust and adaptive regulatory landscape that can keep pace with the dynamic nature of AI and its implications for the financial industry. The goal is not simply to regulate, but to create an environment that promotes responsible innovation and harnesses the transformative potential of AI for the benefit of all.

The burgeoning field of AI-powered investing hinges critically on the availability and processing of vast quantities of financial data. This data, encompassing everything from transactional records and market indicators to personal financial information and investor preferences, is the lifeblood of sophisticated AI algorithms that drive automated trading, portfolio management, and personalized financial advice. However, the collection, use, and storage of such sensitive data are subject to a complex and evolving web of data protection and privacy laws, posing significant challenges and opportunities for fintech companies and financial institutions. Navigating this legal landscape is paramount to ensuring both the responsible deployment of AI in finance and the protection of individual rights.

One of the most influential pieces of legislation globally is the General Data Protection Regulation (GDPR), enacted by the European Union in 2018. The GDPR establishes a comprehensive framework for the processing of personal data, placing significant obligations on organizations that collect, use, or store such data. For AI-powered investing, this translates into strict requirements regarding consent, data minimization, purpose limitation, data security, and individual rights such as the right to access, rectify, erase, and restrict the processing of their personal data. Failing to comply with the GDPR can result in substantial fines, potentially crippling even the most successful fintech startups.

The GDPR's impact extends beyond Europe. Many companies globally adopt GDPR-compliant practices, recognizing the high standard it sets and the potential reputational damage associated with data breaches or non-compliance. The principle of data protection by design and default, a cornerstone of the GDPR, mandates that companies integrate data protection considerations into the design and development of their AI systems from the outset, rather than treating it as an afterthought. This requires a fundamental shift in thinking, embedding privacy considerations into the core architecture and functionality of AI algorithms and data management systems.

Beyond the GDPR, numerous other data protection and privacy laws are relevant to the AI-powered investing landscape. In the United States, the California Consumer Privacy Act (CCPA) and similar state-level laws grant California residents significant rights regarding their personal data, including the right to know what data is collected, the right to delete data, and the right to opt out of the sale of their personal data. These laws are particularly relevant given California's prominent position in the technology industry and its influence on national data protection standards. The potential for a patchwork of state-level laws presents challenges for companies seeking to maintain a unified approach to data privacy across different jurisdictions. The proposed federal privacy legislation seeks to create a more uniform and comprehensive framework across the United States.

Furthermore, sectoral regulations impacting financial institutions, such as the Gramm-Leach-Bliley Act (GLBA) in the United States, add another layer of complexity. The GLBA, designed to protect consumer financial data, mandates specific security and privacy practices for financial institutions, with implications for how AI systems handle personal financial information. Compliance necessitates robust security measures, including encryption, access controls, and regular security audits. The intersection of GLBA and GDPR, for example, necessitates careful consideration of data transfer mechanisms, ensuring compliance with both legislative frameworks when dealing with European citizens' data.

The use of AI in investment decision-making often involves the processing of non-personal data, such as market data, economic indicators, and company financial statements. While not directly subject to the same stringent privacy requirements as personal data, the responsible handling of such information is still crucial to ensure the integrity and reliability of AI-driven investment strategies. Data accuracy, validation, and governance processes are essential for maintaining the trust and confidence of investors. The use of inaccurate or manipulated data can lead to flawed investment decisions, potentially causing significant financial losses and reputational damage. Robust data quality controls are therefore paramount.

The challenge of data anonymization and pseudonymization adds another layer to this complex picture. While these techniques can reduce the risk of identifying specific individuals, they don't eliminate it entirely. Advanced AI algorithms, particularly those leveraging machine learning, may be able to re-identify individuals even from seemingly anonymized datasets. This highlights the need for caution and continuous evaluation of anonymization techniques, particularly in the context of evolving AI capabilities.

The ethical implications are also significant. AI systems trained on biased data can perpetuate and amplify existing inequalities, resulting in discriminatory outcomes. For example, an AI algorithm used in loan applications might unfairly discriminate against certain demographic groups if the training data reflects existing societal biases. To ensure fairness and equity, careful attention must be paid to the composition and quality of training data, along with the implementation of bias detection and mitigation techniques. Transparency and explainability of AI algorithms are also crucial to ensure accountability and allow for scrutiny of decision-making processes.

Data breaches and cybersecurity threats remain a significant concern. The increasing reliance on AI in finance makes institutions more vulnerable to cyberattacks, potentially leading to the exposure of sensitive financial data. Robust cybersecurity measures, including encryption, intrusion detection systems, and regular security audits, are critical to mitigating these risks. Incident response plans should also be in place to quickly address and contain any data breaches. The cost of data breaches can be staggering, including financial losses, legal liabilities, and reputational damage.

International cooperation and harmonization of data protection laws are essential for fostering a global ecosystem for AI-powered investing. Inconsistent regulations across different jurisdictions can create significant challenges for companies operating internationally. A more unified global approach to data protection would not only simplify compliance but also contribute to building greater trust and confidence in the use of AI in finance. This might involve the development of internationally recognized standards and best practices for data privacy and security in the context of AI.

In conclusion, the legal and regulatory landscape surrounding data protection and privacy in AI-powered investing is both complex and dynamic. Navigating this landscape requires a deep understanding of relevant laws and regulations, a commitment to ethical data handling practices, and a proactive approach to risk management and compliance. Failing to adequately address these issues can result in significant legal and financial liabilities, along with reputational damage. However, by embracing responsible data governance and transparent AI practices, fintech companies and financial institutions can unlock the transformative potential of AI while upholding the highest standards of data protection and consumer rights. The continued evolution of both AI and data protection legislation requires ongoing adaptation and collaboration between industry, regulators, and researchers to ensure a safe, secure, and equitable future for AI in finance.

The responsible implementation of AI in financial services necessitates a robust framework of compliance and audit procedures. This isn't simply a matter of ticking boxes; it's a fundamental aspect of ensuring the integrity, security, and ethical operation of AI-driven systems within the highly regulated financial industry. Failure to adhere to these requirements can result in significant penalties, reputational damage, and erosion of investor trust. The complexities involved demand a proactive and comprehensive approach, encompassing meticulous record-keeping, regular audits, and a commitment to continuous improvement.

One crucial element is the meticulous maintenance of comprehensive records detailing all aspects of the AI system's development, deployment, and operation. This includes meticulous documentation of the data sources used, the algorithms employed, the decision-making processes, and any adjustments or modifications made to the system over time. This documentation serves multiple purposes. Firstly, it provides a clear audit trail, enabling regulators and internal auditors to track the system's performance and identify any potential irregularities or biases. Secondly, it fosters transparency and accountability, allowing stakeholders to understand how the AI system functions and the basis for its decisions. Thirdly, it aids in the identification of areas for improvement, enabling the continuous refinement of the AI system to enhance its accuracy, fairness, and efficiency. The format of these records should be standardized, readily accessible, and consistently maintained, employing version control systems to track changes and ensure data integrity. Failure to maintain such records could severely hamper any investigation into potential compliance breaches.

Regular audits are a cornerstone of compliance in the financial industry, and the use of AI necessitates a more sophisticated approach. Traditional audit methodologies may need adaptation to effectively assess the complexities of AI systems. Audits should not only examine the accuracy and reliability of the data used to train and operate the AI but also investigate the algorithms' fairness, transparency, and explainability. Auditors need to assess whether the system's decisions are justifiable, free from bias, and aligned with ethical guidelines and regulatory requirements. This requires expertise in both auditing and AI, potentially involving collaboration between financial auditors and data scientists. The scope of these audits should extend beyond simply reviewing financial reports, encompassing a comprehensive evaluation of the AI system's architecture, data pipelines, decision-making processes, and risk management protocols. The frequency of audits will depend on factors such as the AI system's complexity, its impact on financial operations, and regulatory requirements. Regular, independent, and comprehensive audits are essential to maintaining confidence in the system's integrity and ensuring compliance with relevant regulations.

Specific compliance procedures vary depending on the jurisdiction and the nature of the AI application. However, certain common principles apply across different regulatory environments. Data privacy and security are paramount. AI systems in finance often process sensitive personal and financial data, necessitating robust measures to protect this information from unauthorized access, use, or disclosure. Compliance with regulations such as the GDPR in Europe, the CCPA in California, and other regional or national data protection laws is crucial. This includes obtaining appropriate consent for data processing, implementing data minimization techniques, ensuring data security through encryption and access controls, and providing individuals with rights concerning their personal data, such as the right to access, rectify, or erase their data. Failure to comply with these regulations can result in significant fines and reputational damage.

Furthermore, AI systems must adhere to anti-money laundering (AML) and know-your-customer (KYC) regulations. These regulations require financial institutions to identify and prevent illicit activities such as money laundering and terrorist financing. AI can play a valuable role in enhancing AML/KYC compliance by automating tasks such as transaction monitoring and customer due diligence. However, it's crucial to ensure that these AI systems are designed and implemented in compliance with applicable regulations and that appropriate safeguards are in place to prevent bias or discrimination.

The use of AI in algorithmic trading presents its own set of compliance challenges. Regulations governing high-frequency trading and market manipulation are pertinent here. AI systems must be designed and operated in a manner that prevents market manipulation and ensures fair and orderly markets. Transparency and explainability are essential. Regulators need to understand how AI-driven trading systems make decisions to assess their potential impact on market stability and prevent manipulation.

Model risk management is another critical consideration. AI models, particularly those used in risk assessment and credit scoring, must be carefully validated and monitored to ensure their accuracy and reliability. Regular model validation procedures should be in place to identify and mitigate potential risks associated with model inaccuracies or biases. This involves testing the model's performance against historical and simulated data, evaluating its robustness to different market conditions, and assessing its potential for discriminatory outcomes.

Finally, maintaining a culture of compliance within the organization is crucial. This involves providing employees with adequate training on AI ethics, data privacy, and regulatory compliance, and establishing clear internal policies and procedures. Regular compliance reviews and assessments should be conducted to identify and address potential weaknesses. The commitment must extend from the top management down to ensure that AI is used responsibly and ethically within the organization.

In conclusion, compliance and audit requirements for AI in financial services are multifaceted and require a proactive, comprehensive, and continuously evolving strategy. It's not enough to simply meet minimum regulatory requirements; a commitment to exceeding expectations through robust procedures, ethical considerations, and a culture of transparency is essential. This ensures not only avoidance of legal repercussions but also the fostering of trust with clients, regulators, and the broader financial community, ultimately solidifying a company's position as a responsible and innovative leader in the fintech sector. The continuous evolution of both AI technologies and regulatory landscapes necessitates an ongoing commitment to adaptation, learning, and collaboration to ensure the responsible and ethical deployment of AI in financial services.

The deployment of AI in financial services, particularly within investment strategies, introduces a complex web of liability and accountability. The traditional legal frameworks, designed for human-driven decisions, often struggle to adequately address the intricacies of AI-powered actions. Determining responsibility when an AI system makes an investment decision that results in losses becomes a challenging legal and ethical question. This subsection delves into the legal landscape surrounding liability and accountability in AI-driven investing, examining the potential legal precedents and liabilities involved.

One of the primary challenges lies in identifying the responsible party when an AI system malfunctions or makes a detrimental decision. Is it the developer of the AI algorithm, the financial institution deploying the system, or the individual investor utilizing the AI-powered platform? The answer is rarely straightforward and often depends on the specific circumstances, the contractual agreements in place, and the applicable laws and regulations.

Consider, for example, the case of an AI-powered algorithmic trading system making a series of erroneous trades, resulting in substantial financial losses for the investors. Several parties could potentially be held liable. The developers of the AI algorithm might be held accountable if a design flaw or coding error caused the system to malfunction. This could involve claims of negligence or breach of contract if the AI did not perform as promised. The financial institution deploying the system may also face liability if they failed to adequately test, monitor, or oversee the AI's operation. This could be grounds for claims of negligence or even fraud if the institution knowingly deployed a flawed system. Finally, the investor using the platform might bear some responsibility, particularly if they failed to understand the risks associated with using an AI-powered system or disregarded the warnings provided. This situation exemplifies the multifaceted nature of liability in AI-driven investing.

Existing legal frameworks, such as those governing product liability and negligence, may be partially applicable but often fall short in addressing the unique challenges posed by AI. Traditional product liability laws typically focus on physical products, not software or algorithms. Negligence claims require proving a duty of care, a breach of that duty, causation, and damages. Establishing causation in the context of an AI system's actions can be particularly complex, as the system's decision-making process may be opaque and difficult to decipher. The "black box" nature of some AI algorithms makes it challenging to pinpoint the exact cause of a negative outcome, hindering the ability to definitively establish liability.

Furthermore, the rapid pace of technological innovation in AI often outstrips the ability of legal systems to adapt and provide clear guidelines. Legal precedents are still evolving, and there's a lack of established case law specifically addressing the nuances of AI-driven investment decisions. This uncertainty creates ambiguity and challenges for both developers and users of AI in finance.

To mitigate legal risks, developers and financial institutions deploying AI systems should prioritize transparency and explainability. Creating AI systems that are easily understood and auditable is crucial. This includes documenting the development process, the data used to train the algorithm, and the decision-making logic employed by the system. Such meticulous record-keeping allows for a more thorough investigation in case of disputes or failures, potentially aiding in determining responsibility.

Moreover, robust testing and validation procedures are essential. Thoroughly testing the AI system under various conditions helps identify and rectify potential flaws before deployment. Continuous monitoring of the system's performance after deployment also helps in early detection of anomalies or malfunctions. Implementing effective risk management strategies and incorporating appropriate safeguards into the system's design can minimize the likelihood of errors and mitigate potential liabilities.

Contractual agreements between developers, financial institutions, and investors should explicitly address liability and accountability issues. Clearly defining the roles and responsibilities of each party can help prevent future disputes. This includes detailing the expectations of the AI's performance, the procedures for monitoring and managing risks, and the mechanisms for resolving conflicts arising from system failures. These contracts should also outline the specific liability limitations and insurance coverage relevant to the use of the AI system.

Regulatory bodies are also playing a vital role in shaping the legal landscape surrounding AI in finance. Regulatory frameworks are evolving to address the unique challenges posed by AI-driven systems. These regulations often focus on aspects such as data privacy, security, and algorithmic transparency. Compliance with these regulations is essential for minimizing legal risks. As AI continues to proliferate within the financial industry, we can anticipate further regulatory developments aiming to clarify liability and accountability issues and protect investors from potential harms. These regulations often include provisions for auditing, model validation, and risk management, contributing to the overall objective of enhancing AI's responsible use.

Beyond the legal ramifications, ethical considerations are also paramount. While legal liability focuses on assigning blame and establishing compensation, ethical considerations delve into the broader implications of AI's actions on individuals, markets, and society. The potential for bias in AI algorithms, for example, raises significant ethical concerns, even if it doesn't necessarily translate into legal liability. The responsible development and deployment of AI require careful attention to ethical implications and a commitment to mitigating potential biases. This includes ensuring fairness and promoting inclusivity in the design, implementation, and use of AI systems.

The interplay between legal liability and ethical considerations creates a complex framework guiding the responsible use of AI in finance. Striking a balance between fostering innovation and protecting individuals and markets requires a multi-pronged approach involving robust regulatory frameworks, ethical guidelines, and a commitment to transparency and accountability from all stakeholders. The field is constantly evolving, and the ongoing dialogue between legal professionals, technologists, and ethicists will shape the future of AI in finance. The focus must remain on creating an environment where AI can enhance the financial system while mitigating risks and fostering trust. The path forward hinges on collaboration and proactive measures to establish clearer lines of accountability and responsibility within this rapidly developing landscape. This evolving understanding will inevitably lead to new legal precedents and refined regulatory frameworks in the years to come.

The rapid evolution of artificial intelligence (AI) in financial services necessitates a forward-looking perspective on regulatory trends. Current regulatory frameworks, while addressing some aspects of AI's impact, often lag behind technological advancements. Predicting the precise shape of future regulations is inherently challenging, but analyzing emerging trends and technological developments allows for informed speculation on likely regulatory directions. Several key areas suggest significant changes are on the horizon.

One prominent area is the increasing focus on explainability and transparency in AI systems. The "black box" nature of some AI algorithms, which makes it difficult to understand their decision-making processes, is a major concern for regulators. Future regulations will likely mandate greater transparency, requiring financial institutions to provide clear explanations of how AI systems arrive at their conclusions, particularly when those conclusions impact individuals or markets. This could involve requirements for detailed documentation of algorithms, data sets used for training, and the validation processes employed. Techniques like explainable AI (XAI) will become increasingly crucial, allowing for the interpretation and auditability of AI-driven decisions. Penalties for opaque systems could include limitations on deployment or significant financial penalties. The emphasis on explainability is not merely about accountability; it's also about fostering trust in AI-powered financial systems.

Another critical trend relates to data governance and privacy. AI systems rely heavily on vast quantities of data, raising concerns about data security, privacy, and the potential for bias. Future regulations will likely strengthen data protection laws, extending their scope to encompass the unique challenges presented by AI. This could involve stricter requirements for data anonymization, encryption, and access control. Regulations might also address the use of personal data in training AI algorithms, requiring explicit consent and providing individuals with greater control over their data. Moreover, regulations might focus on mitigating bias in AI algorithms. Algorithms trained on biased data can perpetuate and amplify existing inequalities, leading to unfair or discriminatory outcomes. Future regulations may mandate rigorous assessments for bias in AI systems and require mitigation strategies to ensure fairness and equity. This includes establishing mechanisms for auditing and detecting bias, demanding transparency in the data used for training, and imposing penalties for systems exhibiting discriminatory tendencies.

The liability framework surrounding AI in finance is another area ripe for change. As discussed earlier, the challenges in determining responsibility when an AI system malfunctions or makes a detrimental decision are significant. Future regulations will likely clarify liability, specifying the roles and responsibilities of developers, financial institutions, and investors. This could involve establishing a clearer chain of accountability, defining standards of care for AI development and deployment, and creating mechanisms for resolving disputes arising from AI-related errors. The establishment of specific legal entities, potentially insurance schemes, to handle AI-related liabilities may be another significant development. This dedicated framework could improve the efficiency and fairness of handling disputes, providing a mechanism for compensation to those affected by AI malfunctions.

Furthermore, regulatory bodies are likely to increasingly focus on the systemic risks posed by AI in finance. The interconnectedness of financial markets means that a failure in one AI system could have cascading effects, impacting multiple institutions and the broader financial system. Future regulations will likely address these systemic risks by introducing stricter oversight, stress testing requirements, and contingency planning for AI-driven systems. This could include establishing regulatory sandboxes, controlled environments for testing and evaluating AI systems before full deployment, to mitigate potential systemic risks. This approach combines the need for innovation with the imperative of avoiding widespread financial disruptions. More emphasis on the resilience of algorithms and infrastructural robustness will be crucial, necessitating extensive stress tests and robust cybersecurity measures.

The increasing use of AI in areas like algorithmic trading and credit scoring is driving the need for more sophisticated regulatory frameworks. These frameworks might require more frequent and rigorous audits of AI systems, possibly incorporating independent validation by third-party experts. This external scrutiny aims to ensure adherence to regulatory standards and prevent the deployment of flawed or biased systems. The need for standardized methodologies for validating algorithms will likely be a focal point. These standards would ensure uniformity in testing procedures and create a basis for comparing and assessing the risk of different AI systems.

The international harmonization of AI regulations is also likely to be a major trend. The global nature of financial markets necessitates a coordinated approach to regulating AI, preventing regulatory arbitrage and ensuring a level playing field. International cooperation among regulatory bodies will be crucial in developing consistent standards and promoting cross-border collaboration in the supervision of AI systems. This collaboration could involve shared information, joint investigations, and the development of common regulatory frameworks. A global approach ensures a unified response to the challenges presented by AI in finance, avoiding conflicts and fostering a consistent regulatory environment.

The impact of these future regulatory trends will be significant. Financial institutions will need to invest in technology and expertise to comply with stricter requirements for explainability, data governance, and risk management. The cost of compliance could be substantial, potentially influencing the adoption and deployment of AI systems. However, robust regulatory frameworks are also likely to increase investor confidence and trust in AI-powered financial services. This increased trust could lead to greater adoption and broader benefits, driving the overall growth of the AI sector in finance. The regulatory landscape will continue to evolve as technology advances, creating a dynamic interaction between innovation and regulation. The key lies in finding a balance, allowing innovation to flourish while safeguarding the stability and integrity of the financial system. This requires a collaborative effort between regulators, financial institutions, and technology developers, fostering open communication and a shared understanding of the challenges and opportunities presented by AI in finance. The ultimate goal is to harness the power of AI to improve financial services while mitigating risks and protecting consumers and the financial system as a whole. The journey towards this goal will require continuous adaptation and a willingness to embrace the transformative potential of AI while addressing its inherent challenges.

This exercise will walk you through constructing a simplified AI-driven investment portfolio. We'll leverage readily available tools and publicly accessible data to illustrate the core concepts. This isn't intended to be a comprehensive, ready-to-deploy investment strategy, but rather a pedagogical tool to demonstrate the integration of AI in portfolio management. Remember, any investment strategy carries risk, and past performance is not indicative of future results. Consult with a qualified financial advisor before making any investment decisions.

Step 1: Data Acquisition and Preparation

Our first step involves gathering relevant financial data. For this exercise, we'll focus on a small selection of publicly traded stocks. You can obtain this data from various sources, including financial APIs (like Alpha Vantage, Tiingo, or IEX Cloud), or by scraping data from websites like Yahoo Finance (though scraping requires careful consideration of website terms of service). We'll need historical price data (at least the adjusted closing price) for a period of several years for each selected stock. The length of the historical period influences the robustness of the model; longer periods usually provide better results but also require more computational resources. For this exercise, we'll use three years of daily data.

Let's assume we've chosen Apple (AAPL), Microsoft (MSFT), and Google (GOOGL). Once the data is downloaded, it needs to be cleaned and preprocessed. This involves handling missing values (if any), converting data types appropriately (dates to datetime objects), and potentially normalizing or standardizing the data to improve model performance. This data cleaning is crucial; inaccurate or inconsistent data will lead to unreliable model outputs. Most programming languages (Python, R) offer robust libraries (like pandas in Python) for handling data manipulation and cleaning.

Step 2: Feature Engineering

Raw price data alone might not be sufficient for effective AI-driven portfolio construction. We need to engineer relevant features that could influence future stock performance. Common features include:

Moving Averages: Calculate various moving averages (e.g., 50-day, 200-day) to smooth out price fluctuations and identify trends.

Relative Strength Index (RSI): This momentum indicator measures the magnitude of recent price changes to evaluate overbought or oversold conditions.

Bollinger Bands: These bands plot standard deviations around a moving average, providing a visual representation of volatility.

Volume: Trading volume can provide insights into market sentiment and potential price movements.

Returns: Calculate daily, weekly, or monthly returns to quantify the performance of each stock.

These are just a few examples; other features, potentially derived from more advanced technical analysis indicators or fundamental data (earnings reports, etc.), could be incorporated. The choice of features significantly impacts model accuracy. Experimentation is key.

Step 3: Model Selection and Training

Now that we have our prepared dataset with engineered features, we can select a suitable machine learning model. For portfolio optimization, reinforcement learning (RL) algorithms are often preferred. However, for this simplified exercise, we'll utilize a simpler supervised learning approach – a regression model. Specifically, we'll employ a linear regression model to predict future returns based on the engineered features. This choice allows for ease of understanding and implementation. More complex models, such as support vector regression (SVR) or gradient boosting machines (GBM), could also be explored for potentially better performance, but increase complexity.

Before training our model, we need to split our data into training and testing sets. The training set is used to train the model, and the testing set evaluates its performance on unseen data. A common split is 80% training and 20% testing. We then train the linear regression model using the training data. Libraries like scikit-learn in Python provide convenient tools for this. The process involves fitting the model to the historical data, which allows the algorithm to learn the relationship between the input features (moving averages, RSI, etc.) and the target variable (future returns).

Step 4: Portfolio Optimization

Once the model is trained and validated on the testing set, we can use it to predict future returns for our chosen stocks. Based on these predictions, we can construct a portfolio. A simple optimization approach would be to allocate capital proportionally to the predicted returns. For example, if the model predicts a 10% return for AAPL, 5% for MSFT, and 2% for GOOGL, we'd allocate a larger portion of our capital to AAPL.

However, this simple approach doesn't consider risk. A more sophisticated approach would involve incorporating risk measures (like standard deviation or variance) into the optimization process. This could be achieved using techniques like the Markowitz mean-variance optimization, which balances expected return and risk. Modern portfolio theory (MPT) provides a robust framework for this task. Libraries like `cvxopt` in Python can be used for solving quadratic programming problems which are inherent in MPT.

Step 5: Backtesting and Evaluation

After determining portfolio weights, we perform backtesting. This involves simulating the portfolio's performance using historical data. We can use the model's predictions on the historical test set to assess how well it would have performed in the past. Key metrics include Sharpe Ratio, Sortino Ratio, and maximum drawdown to evaluate risk-adjusted returns. These metrics help in evaluating the overall performance of our AI-driven strategy.

It's important to remember that backtesting provides only an approximation of future performance. Market conditions constantly evolve, and any model's effectiveness can diminish over time. The backtesting results should be interpreted cautiously and not taken as a guarantee of future success. Regular re-evaluation and model retraining are vital for maintaining the accuracy and effectiveness of the AI-driven portfolio management strategy.

Step 6: Refinement and Iteration

The process of building an AI-driven portfolio is iterative. After backtesting, you might identify areas for improvement. This could involve trying different models, adding or removing features, adjusting the optimization parameters, or using more sophisticated risk management techniques. The key is to experiment, learn from the results, and continuously refine the strategy to improve its performance.

For instance, you could consider incorporating sentiment analysis from news articles or social media to supplement your quantitative data. Sentiment analysis can provide insights into market sentiment and may help predict future price movements. You could also explore the inclusion of fundamental data to enhance your model. This would involve gathering and preprocessing information such as earnings per share (EPS), price-to-earnings ratio (P/E), and debt-to-equity ratios.

Another refinement could involve employing techniques to handle model overfitting, a common problem in machine learning where the model performs exceptionally well on the training data but poorly on unseen data. Regularization techniques, such as L1 or L2 regularization, can help mitigate this issue. Furthermore, you may want to investigate more advanced portfolio optimization strategies that consider transaction costs and tax implications, further enhancing the realism of your simulation.

Remember, this exercise provides a simplified illustration. Real-world AI-driven portfolio management is substantially more complex and involves numerous considerations that extend beyond the scope of this simplified example. However, this exercise establishes a foundational understanding of the core concepts and the practical implementation process.

This exercise delves into the realm of sentiment analysis, a powerful technique for gauging market sentiment from textual data. We'll focus on applying sentiment analysis to financial news articles, leveraging readily available tools and APIs to extract insights that can potentially inform investment decisions. This is not a foolproof method for predicting market movements, but rather a complementary tool to enhance your existing analytical framework. Remember that market dynamics are complex and influenced by numerous factors beyond the scope of sentiment alone. Always exercise caution and conduct thorough due diligence before making any investment decisions.

Step 1: Data Acquisition and Selection

The first hurdle is acquiring a suitable dataset of financial news articles. Several sources provide this data, albeit with varying levels of access and cost. Free sources include scraping news websites (again, always respecting terms of service), but this approach can be time-consuming and requires significant data cleaning due to inconsistencies in website structure and formatting. More reliable, although often paid, options include specialized financial news APIs that provide structured data feeds, minimizing preprocessing effort. Examples include those offered by companies like Dow Jones, Thomson Reuters, or FactSet. For this exercise, let's assume you've selected an API or gathered a sample dataset of news articles concerning a specific company or sector. The articles should cover a substantial period to provide a statistically significant dataset for analysis. For demonstration purposes, we will assume we have collected 1000 news articles related to Apple (AAPL) over the past year.

The selection of news sources is crucial. Consider the bias and credibility of each source. Articles from reputable financial news outlets generally yield more reliable sentiment scores compared to those from less established sources or social media platforms. Ideally, your dataset should encompass a diversity of viewpoints, ensuring a balanced representation of market sentiment. Prioritize articles directly discussing the company's performance, financial announcements, or related market events. Articles that are tangentially related or focused on broader economic trends may not significantly contribute to the sentiment analysis of a specific company.

Step 2: Data Preprocessing and Cleaning

Before sentiment analysis, the data requires careful preprocessing. This often involves several steps:

Data Cleaning: This involves removing irrelevant characters, such as HTML tags (if scraping from websites), correcting spelling errors, and handling missing values. Missing values could be handled by simply removing articles with missing information or, if appropriate, by imputation methods, although imputation introduces potential biases.

Text Normalization: This converts text to a standardized format. Common techniques include lowercasing all text, removing punctuation, and stemming or lemmatization (reducing words to their root forms). Consider the implications of each method; stemming might lose some nuanced meanings, whereas lemmatization retains more information.

Tokenization: This step breaks the text into individual words or phrases (tokens). This is a fundamental step for many natural language processing (NLP) tasks. The choice of tokenizer can impact the results. Simple tokenizers split on whitespace, while more advanced tokenizers can handle punctuation and other linguistic complexities.

Stop Word Removal: Stop words are common words (like "the," "a," "is") that often don't contribute much to sentiment analysis. Removing them reduces computational overhead and improves accuracy.

The choice of tools and techniques greatly impacts efficiency and accuracy. Python offers several libraries, including `NLTK`, `spaCy`, and `transformers` that simplify these steps.

Step 3: Sentiment Analysis Techniques

Several techniques perform sentiment analysis. These range from simple rule-based approaches to sophisticated machine learning models.

Lexicon-based approaches: These utilize pre-built dictionaries (lexicons) that assign sentiment scores to words (e.g., "good" = positive, "bad" = negative). These are relatively simple but can be limited as the subtleties of language are not always captured. VADER (Valence Aware Dictionary and sEntiment Reasoner) is a popular lexicon-based tool specifically designed for analyzing social media text, and it might be well suited for analyzing news headlines or shorter financial news snippets.

Machine learning approaches: These train models on labeled data (news articles with pre-assigned sentiment scores) to learn patterns and predict sentiment in new text. These are often more accurate than lexicon-based approaches, but require a substantial labeled training dataset. You can use pre-trained models available through platforms such as Hugging Face. These models are often fine-tuned on large corpora of text and provide strong baseline performance. Fine-tuning these models on a dataset of financial news might lead to enhanced performance specifically relevant to financial language and terminology.

Hybrid approaches: Combine lexicon-based and machine learning techniques to leverage the strengths of both. This strategy can improve the accuracy and robustness of sentiment analysis.

For this exercise, let's use a pre-trained model from Hugging Face. These models are computationally efficient and readily accessible.

Step 4: Implementation and Analysis

Let's walk through a simplified Python example using a pre-trained sentiment analysis model and the Hugging Face Transformers library:

```python
from transformers import pipeline
classifier = pipeline("sentiment-analysis")
text = "Apple's earnings exceeded expectations, leading to a surge in stock price."
```

```
result = classifier(text)
print(result) Output: [{'label': 'POSITIVE', 'score': 0.98}]
```

This code snippet utilizes a pre-trained sentiment analysis pipeline. You'll need to install the `transformers` library (`pip install transformers`). Substitute `text` with the content of each news article from your dataset. The output provides a label ('POSITIVE', 'NEGATIVE', or 'NEUTRAL') and a corresponding score indicating the confidence of the classification. You can iterate through your entire dataset to obtain sentiment scores for all articles.

Once you've obtained the sentiment scores, you can analyze them using various statistical methods. Calculate the average sentiment score for a given period, observe trends in sentiment over time, or analyze the correlation between sentiment and the company's stock performance. Visualization techniques like line graphs can illustrate how sentiment evolves alongside stock price movements, offering valuable insights into the relationship between news sentiment and market fluctuations. Remember to consider potential biases in your analysis.

Step 5: Interpreting Results and Limitations

Interpreting sentiment analysis results requires caution. A high positive sentiment doesn't guarantee a stock price increase, nor does negative sentiment automatically predict a decline. Many other factors influence stock prices. Sentiment analysis provides one piece of the puzzle, not the complete picture. Consider the following caveats:

Context is crucial: Sentiment analysis might misinterpret sarcasm or irony. The context surrounding words and phrases significantly impacts their sentiment.

Nuance and subjectivity: Human language is complex, and some aspects might not be accurately captured by algorithms. The same sentence can have different connotations depending on the audience.

Data limitations: The accuracy of sentiment analysis depends heavily on the quality and representativeness of the input data. A biased dataset will lead to biased results.

Overfitting: Overreliance on sentiment analysis without considering other financial indicators can lead to inaccurate predictions and poor investment decisions.

This exercise illustrates the application of sentiment analysis in a financial context. It showcases the potential for enhancing investment strategies with AI-driven techniques, but also highlights the crucial need for careful interpretation and the integration of multiple analytical tools for robust investment decision-making. Combine sentiment analysis with fundamental and technical analysis to gain a more holistic perspective. Always remember to incorporate risk management principles into your overall investment strategy.

This exercise focuses on evaluating the historical performance of an algorithmic trading strategy. We'll use publicly available data and common performance metrics to assess its effectiveness. The strategy we'll evaluate is a simple mean reversion strategy applied to the daily prices of the S&P 500 index. Mean reversion strategies are based on the assumption that asset prices tend to revert to their average value over time. This strategy will identify potential trading opportunities when the price deviates significantly from its moving average.

Step 1: Data Acquisition and Preparation

First, we need historical daily price data for the S&P 500. Several sources provide this freely, including Yahoo Finance, Google Finance, and financial data APIs like Alpha Vantage. For this exercise, let's assume you've downloaded the daily adjusted closing prices for the S&P 500 for the past five years. You can typically download this data in CSV format. The adjusted closing price accounts for corporate actions like dividends and stock splits, ensuring a consistent price series for backtesting. Import this data into a spreadsheet program like Excel or a programming language like Python using libraries such as Pandas.

Data cleaning is essential. Inspect your dataset for missing values or inconsistencies. Missing data can be handled through several methods, such as linear interpolation (estimating missing values based on neighboring data points) or simply removing the affected rows. Inconsistencies might require manual correction or removal of problematic data points. For this exercise, we'll assume a clean dataset with no missing values. Once cleaned, organize your data with a date column and a corresponding adjusted closing price column.

Step 2: Defining the Algorithmic Trading Strategy

Our mean reversion strategy will involve the following steps:

1.

Calculate the moving average: We will use a 20-day simple moving average (SMA) as a benchmark. The SMA is simply the average of the closing prices over the past 20 days. The choice of the moving average window (20 days in this case) is crucial; different window sizes will result in different trading signals and performance. Experimentation with different window sizes can help optimize the strategy.

2.

Generate trading signals: A buy signal is generated when the current price falls below the 20-day SMA by a certain percentage (let's say 1%). Conversely, a sell signal is generated when the current price rises above the 20-day SMA by the same percentage. This percentage threshold acts as a buffer to prevent over-trading based on minor fluctuations.

3.

Manage trades: This strategy will employ a simple position sizing approach, allocating a fixed percentage of the portfolio value to each trade. For example, we could allocate 10% of the portfolio value to each buy or sell order. Stop-loss orders are not explicitly included in this basic strategy for simplicity but are highly recommended for real-world applications to mitigate risks.

4.

Transaction Costs: In a real-world scenario, we must account for transaction costs such as brokerage fees and commissions. These costs can significantly impact the overall profitability of a trading strategy. For this exercise, we'll initially ignore transaction costs for simplicity but add this factor in a later analysis.

Step 3: Backtesting the Strategy

Backtesting involves applying the trading strategy to historical data to assess its potential performance. We'll use the historical S&P 500 data and the steps defined in Step 2.

Here's a simplified Python example using Pandas and NumPy to backtest the strategy:

```python
import pandas as pd
import numpy as np
```

Load the S&P 500 data

data = pd.read*csv("sp500*data.csv", index*col="Date", parse*dates=True)

Calculate the 20-day SMA

```
data["SMA"] = data["Adj Close"].rolling(window=20).mean()
```

Generate trading signals

```
data["Signal"] = 0.0
    data["Signal"][data["Adj Close"] < data["SMA"] 0.99] = 1.0 Buy signal
    data["Signal"][data["Adj Close"] > data["SMA"] 1.01] = -1.0 Sell signal
```

Calculate daily returns

```
data["Returns"] = data["Adj Close"].pct_change()
```

Calculate strategy returns (simplified, ignoring transaction costs and position sizing)

data["Strategy Returns"] = data["Signal"].shift(1) data["Returns"]

Calculate cumulative returns

```
data["Cumulative Returns"] = (1 + data["Strategy Returns"]).cumprod()
data["Cumulative Buy and Hold"] = (1 + data["Returns"]).cumprod()
```

Analyze results

print(data[["Adj Close", "SMA", "Signal", "Strategy Returns", "Cumulative Returns", "Cumulative Buy and Hold"]])
```
```

This code provides a basic framework. Adapt it based on your data and preferred programming environment. The "Cumulative Returns" column will show the overall performance of the algorithmic trading strategy. The "Cumulative Buy and Hold" will show what the return would have been if one simply held the S&P 500. This provides a crucial benchmark for comparison.

Step 4: Performance Evaluation Metrics

Once backtested, evaluate the strategy using various metrics:

Annualized Return: The average annual return generated by the strategy. This indicates the overall profitability.

Sharpe Ratio: A risk-adjusted return metric, measuring the excess return per unit of risk. A higher Sharpe ratio suggests better risk-adjusted performance.

Maximum Drawdown: The largest percentage decline from a peak to a trough in the cumulative returns. This metric highlights the downside risk of the strategy.

Sortino Ratio: Similar to the Sharpe ratio, but only considers downside deviations, giving a more accurate picture of risk-adjusted returns if negative returns are more problematic.

Calmar Ratio: Measures the annualized return relative to the maximum drawdown. This metric emphasizes risk-adjusted return and downside protection.

Win Rate: The percentage of trades that resulted in profits. This shows the consistency of the strategy's winning trades.

Average Trade Length: The average number of days a position is held before being closed. This can offer insights into the strategy's trading frequency and market exposure.

Transaction Costs: After the initial analysis, incorporate realistic transaction costs into your backtesting. Re-evaluate the performance metrics to determine the impact of these costs.

Step 5: Refinement and Optimization

The initial backtest provides a starting point. Experiment with different parameters (moving average window, percentage thresholds, position sizing) to optimize the strategy. Consider more sophisticated risk management techniques like stop-loss orders and trailing stops. Explore alternative indicators or combine this mean reversion strategy with other approaches for potentially improved performance. Always be mindful of overfitting. Over-optimizing a strategy to historical data may lead to poor future performance. Out-of-sample testing (applying the strategy to data not used in its development) is crucial for validating the robustness of the optimized strategy.

Step 6: Conclusion and Further Exploration

This exercise demonstrates the fundamental process of evaluating an algorithmic trading strategy. Remember that backtesting results aren't guarantees of future performance. Market conditions change, and strategies that performed well historically may not perform as well in the future. Thorough analysis, careful interpretation of results, and robust risk management are essential for success in algorithmic trading. Further exploration might involve incorporating machine learning techniques to predict future price movements or using more advanced statistical models to refine your risk assessment. Always conduct thorough research and exercise caution before implementing any algorithmic trading strategy with real capital. The world of algorithmic trading involves continuous learning and adaptation to market dynamics.

This exercise delves into the critical area of algorithmic bias within AI-driven investment models. Algorithmic bias, stemming from flawed data or biased algorithms, can lead to unfair or discriminatory outcomes, impacting investment decisions and potentially generating significant financial losses. This exercise will walk you through a step-by-step process of identifying and mitigating such biases in a hypothetical scenario involving a credit scoring model.

Step 1: Understanding the Hypothetical Scenario

Let's consider a simplified AI-driven credit scoring model used by a lending institution. This model uses historical loan application data to predict the creditworthiness of applicants and subsequently determines loan approval or rejection. The data includes features such as age, income, employment history, credit history (including past defaults), and address. The model's objective is to minimize loan defaults while maximizing the number of approved loans.

Step 2: Data Acquisition and Exploration

Before analyzing for bias, we need a dataset representative of the lending institution's historical loan applications. Ideally, this dataset should be extensive and span several years to capture various economic conditions and demographic trends. The dataset must be carefully examined for missing values, inconsistencies, and outliers. Data cleaning is crucial to prevent biased results. Methods like imputation (filling in missing values) or removal of outliers should be applied judiciously, documenting all such decisions.

Following data cleaning, exploratory data analysis (EDA) becomes essential. EDA involves visualizing the data through histograms, scatter plots, and other visual tools to understand the distribution of each feature and identify potential correlations. This step allows us to observe if certain demographics are under-represented or over-represented in the dataset. For instance, are certain age groups, income brackets, or geographical locations significantly under-represented, which could lead to biased model predictions for those groups?

Step 3: Identifying Potential Biases

Once the data is cleaned and explored, we can actively look for potential biases. Several techniques can be employed:

Statistical Analysis: Calculate the distribution of key features for different demographic groups. Perform statistical tests (like t-tests or chi-squared tests) to determine if there are statistically significant differences in loan approval rates or default rates across these groups. Are there statistically significant differences in approval rates between men and women with similar credit scores and incomes? This would highlight potential gender bias.

Visual Inspection: Create visualizations (e.g., bar charts or box plots) to compare the distribution of features and outcomes (loan approval/rejection, default/non-default) across different demographic subgroups. Visualizing the data can quickly reveal disparities that might not be immediately apparent through statistical analysis alone. For example, a box plot comparing income distribution for approved and rejected loan applications might reveal a significant difference in the median income required for approval across different racial groups.

Fairness Metrics: Employ fairness metrics specifically designed to assess algorithmic bias. These metrics quantify the extent of disparity in outcomes across different protected groups. Common metrics include:

Disparate Impact: Measures the ratio of the positive outcome rate (loan approval) for the favored group to the positive outcome rate for the disadvantaged group. A disparate impact ratio significantly greater than 1 indicates potential bias.

Equal Opportunity: Compares the true positive rate (correctly predicting loan repayment) for different groups. Significant differences indicate unequal opportunities.

Predictive Rate Parity: Examines whether the model's predictions are equally accurate for all groups. Significant differences suggest bias.

Step 4: Mitigating Algorithmic Bias

Once potential biases are identified, several strategies can be employed to mitigate them:

Data Preprocessing: Address imbalances in the dataset. Techniques include:

Oversampling: Increase the representation of under-represented groups by duplicating data points.

Undersampling: Reduce the representation of over-represented groups by removing data points.

SMOTE (Synthetic Minority Oversampling Technique): Generate synthetic data points for under-represented groups to balance the dataset. This creates new data points based on the characteristics of existing data points in the minority class.

Algorithmic Modifications: Employ algorithms less susceptible to bias:

Fairness-Aware Algorithms: These algorithms are designed to incorporate fairness constraints during the model training process. They explicitly aim to minimize disparities in outcomes across protected groups.

Regularization Techniques: Techniques like L1 or L2 regularization can help prevent overfitting to biased data and improve the generalization ability of the model. This reduces the model's reliance on spurious correlations that might perpetuate bias.

Feature Engineering: Create new features that address biases. For example, if income is found to be biased, consider using income relative to the cost of living in the applicant's region to adjust for regional variations. This nuanced approach may improve the model's fairness.

Post-Processing: Adjust the model's predictions after training to mitigate bias. This involves manipulating the model's output scores to reduce disparities in outcomes across protected groups, but it needs careful consideration to avoid impacting overall model accuracy.

Step 5: Model Re-evaluation and Iteration

After implementing bias mitigation techniques, it's crucial to re-evaluate the model's performance and fairness. This involves repeating Step 3, using the same fairness metrics to assess if the implemented techniques have effectively reduced bias. It's an iterative process; you may need to refine your approach based on the re-evaluation. Remember that completely eliminating bias might be impossible, but the goal is to minimize it to acceptable levels while maintaining the model's predictive accuracy.

Step 6: Documentation and Transparency

Throughout this process, maintain meticulous documentation. Record all data preprocessing steps, algorithmic modifications, and fairness metric results. Transparency is vital for building trust and ensuring accountability. Communicate the model's limitations and potential biases to stakeholders, highlighting the efforts taken to mitigate those biases.

This exercise highlights the importance of proactive bias detection and mitigation in AI-driven investment models. By systematically addressing biases, we can improve the fairness, equity, and ultimately the overall effectiveness of such models. The ongoing evolution of fairness-aware algorithms and techniques emphasizes the importance of continued learning and adaptation within the field of AI in finance. Remember, ethical considerations are paramount and should always guide the development and deployment of AI-powered systems in the financial domain. Failing to address bias not only risks financial losses but also erodes trust and creates systemic inequalities.

This exercise focuses on creating a robust personal risk management plan specifically tailored to your AI-driven investment strategy. Unlike traditional investment strategies, incorporating AI introduces new layers of complexity and potential risks. Therefore, a comprehensive plan must account for these unique challenges. This plan isn't a static document; it should be a living, breathing strategy, reviewed and adjusted regularly to reflect market changes, technological advancements, and your evolving risk tolerance.

The first step involves a thorough self-assessment. Begin by honestly evaluating your current financial situation. This includes listing all your assets (cash, stocks, bonds, real estate, etc.), liabilities (debts, loans, mortgages, etc.), and income sources. A detailed inventory is crucial. Consider using spreadsheets or financial planning software to streamline this process. Accuracy is paramount, as this forms the bedrock of your risk management plan. Underestimating assets or overestimating liabilities will lead to a flawed plan.

Next, define your investment goals. Are you saving for retirement, a down payment on a house, your children's education, or simply wealth accumulation? Be specific. Instead of broadly stating "retirement," specify a desired retirement income and target date. This clarity helps determine your investment timeline and risk tolerance. Ambiguous goals lead to poorly defined strategies and potentially unsuitable investments.

The next crucial aspect is assessing your risk tolerance. This involves understanding your emotional response to potential losses. Are you comfortable with significant market fluctuations, or do you prefer a more conservative approach that prioritizes capital preservation? This isn't solely a matter of your age or financial situation. Psychological factors play a significant role. Some individuals, regardless of their wealth, are inherently risk-averse, while others are comfortable with higher risk for potentially higher returns. Several online questionnaires can help you gauge your risk tolerance, but honest self-reflection is equally crucial.

Once you understand your risk tolerance, you can begin classifying your investments according to their risk profiles. AI-driven investment strategies often involve exposure to complex algorithms and data-driven models. Therefore, understanding the risks inherent in these tools is vital. This includes identifying potential vulnerabilities such as algorithmic bias, data errors, or unexpected market reactions to AI-driven trading. You might consider categorizing your investments based on factors such as the algorithm's complexity, the historical performance of the underlying model, and the level of human oversight involved in the investment process.

Consider diversifying your investment portfolio across different asset classes and AI-driven strategies. Over-reliance on a single algorithm or investment approach creates undue exposure to the risks specific to that approach. Diversification is a time-tested principle that minimizes the impact of potential losses in any one area. This isn't merely about spreading your investments across different stocks and bonds; it also entails diversifying the AI models and strategies you employ. This might involve using multiple algorithms with different underlying data sets and methodologies, or combining AI-driven strategies with traditional investment approaches.

Thorough due diligence is essential when selecting AI-driven investment tools. Examine the track record of the algorithm, its developers' expertise, and the data used to train the model. Scrutinize independent reviews and assessments. Understand the algorithm's limitations and potential biases. Transparency is vital. Avoid black-box algorithms where the decision-making process is opaque and cannot be readily scrutinized. The more transparent and explainable the AI model, the better you can understand and manage its inherent risks.

Developing contingency plans is paramount. What will you do if the market experiences a significant downturn? How will you handle unexpected losses? Having well-defined strategies for various scenarios minimizes the emotional impact of unforeseen circumstances. This could involve adjusting your investment strategy, rebalancing your portfolio, or drawing upon emergency funds. These plans should be documented and reviewed regularly to ensure they remain relevant.

Regular monitoring and evaluation are crucial. Track the performance of your AI-driven investments, comparing them against your defined goals and risk tolerance. Use key performance indicators (KPIs) appropriate for your investment strategy. These could include Sharpe ratios, Sortino ratios, maximum drawdown, and other risk-adjusted performance metrics relevant to your AI-driven approach. Regular review allows for timely adjustments and helps prevent significant losses.

Finally, consider professional advice. A financial advisor specializing in AI-driven investment strategies can provide valuable insights and support. They can help you navigate the complexities of these tools, identify potential risks, and adjust your plan to reflect market conditions and your changing circumstances. While this isn't mandatory, seeking expert guidance can be invaluable, especially when dealing with sophisticated AI investment tools.

This personalized risk management plan is a framework. Adapt it to your specific circumstances. Regular review and refinement are key. The constantly evolving landscape of AI in finance necessitates an adaptive and proactive approach to risk management. By following this process, you can significantly increase the likelihood of achieving your investment goals while minimizing potential losses associated with AI-driven strategies. Remember, while AI offers exciting opportunities, understanding and mitigating the inherent risks remains crucial for successful investing.

This book has explored the transformative potential of artificial intelligence (AI) in the realm of finance, examining its capabilities and limitations across various applications. We've journeyed from the foundational concepts of AI and machine learning to their practical implementations in investment strategies, risk management, and regulatory frameworks. The overarching theme has been the symbiotic relationship between human expertise and AI's computational power – a partnership that, when carefully managed, can unlock significant opportunities while mitigating inherent risks.

Let's revisit some of the key concepts that have formed the backbone of our exploration. Firstly, we established the fundamental differences between traditional financial modeling and AI-powered approaches. Traditional methods, often reliant on historical data and established statistical models, possess inherent limitations in predicting unforeseen market events or adapting to rapid changes in market dynamics. AI, with its capacity for processing vast datasets and identifying complex patterns often invisible to human analysts, offers a powerful alternative. However, this power comes with responsibilities. The "black box" nature of some AI algorithms, where the decision-making process lacks transparency, presents a critical challenge. We emphasized the importance of explainable AI (XAI) – models that provide insights into their decision-making processes, allowing for greater understanding, trust, and effective risk management.

The ethical considerations surrounding AI in finance were also central to our discussion. Algorithmic bias, a pervasive issue stemming from biased training data, can lead to unfair or discriminatory outcomes. We explored methods for mitigating bias, including careful data curation, rigorous testing, and ongoing monitoring of AI systems for any signs of discriminatory behavior. The potential for AI-driven systems to exacerbate existing inequalities or create new ones underscores the need for responsible development and deployment of these technologies. Furthermore, the potential for market manipulation through sophisticated AI-driven trading strategies necessitates robust regulatory frameworks to maintain market integrity and protect investors.

We delved into the various applications of AI in finance, starting with algorithmic trading. We explored how AI algorithms can analyze vast datasets to identify profitable trading opportunities, executing trades at speeds and scales impossible for human traders. However, the inherent risks associated with high-frequency trading, including the potential for cascading failures and market instability, were highlighted. We also examined the role of AI in portfolio management, where AI can optimize portfolio construction based on individual investor preferences, risk tolerance, and financial goals. The ability of AI to personalize investment strategies and adapt dynamically to changing market conditions represents a significant advancement.

Risk management, as a critical component of any successful financial strategy, took center stage in our analysis. The integration of AI into risk management systems allows for more accurate and timely identification of potential risks, enabling proactive mitigation strategies. However, the complexity of AI algorithms and the potential for unexpected failures necessitate robust risk management frameworks, incorporating both traditional risk assessment techniques and AI-specific safeguards. We emphasized the importance of understanding the limitations of AI systems and developing contingency plans to address unforeseen events. A thorough understanding of the algorithm's methodology, its training data, and its limitations is crucial for managing risk effectively. This requires not only technical expertise but also an understanding of the broader financial context and potential systemic risks.

The regulatory landscape surrounding AI in finance is rapidly evolving. We discussed the challenges faced by regulators in keeping pace with the rapid advancements in AI technology and the need for frameworks that promote innovation while protecting investors and maintaining financial stability. The development of clear guidelines for the use of AI in financial services, addressing issues such as transparency, accountability, and bias, is critical. Collaboration between industry stakeholders, regulators, and researchers is essential for developing robust and effective regulatory frameworks. The goal is not to stifle innovation but to ensure its responsible and ethical application.

Throughout this book, we've stressed the importance of a human-centered approach to AI in finance. AI should be viewed not as a replacement for human expertise but as a powerful tool to augment human capabilities. The synergy between human judgment and AI's analytical prowess is key to achieving optimal outcomes. Humans bring critical thinking, ethical judgment, and an understanding of the broader context, while AI offers speed, scalability, and the ability to identify patterns that may escape human observation. This collaborative approach ensures responsible innovation, minimizing risks, and maximizing the benefits of AI in finance.

The future of AI in finance is brimming with potential. Advancements in machine learning, natural language processing, and other AI technologies are continuously expanding the possibilities. We can anticipate further advancements in algorithmic trading, personalized investment strategies, fraud detection, and regulatory compliance. However, these advancements must be accompanied by a commitment to ethical considerations, robust risk management, and effective regulation. The successful integration of AI in finance hinges on a careful balance between technological innovation and responsible governance.

Looking ahead, the continued development of XAI techniques will be crucial in fostering trust and transparency in AI-driven financial systems. The ability to understand how AI models arrive at their decisions is essential for effective oversight and risk management. Furthermore, the development of more sophisticated methods for detecting and mitigating algorithmic bias will be critical in ensuring fairness and equity in financial markets.

The role of human expertise remains paramount. While AI can automate many tasks, human judgment and oversight are indispensable in navigating the complexities of the financial world. The development of hybrid models, combining the strengths of human expertise with the capabilities of AI, will be a key trend in the future. The successful integration of AI in finance requires a skilled workforce capable of understanding and managing these powerful tools. This underscores the need for education and training programs to equip professionals with the skills needed to navigate the changing landscape of AI in finance.

The evolution of regulatory frameworks will also be a defining factor in shaping the future of AI in finance. Regulators face the challenge of creating a regulatory environment that fosters innovation while safeguarding against potential risks. The development of adaptable and forward-looking regulatory frameworks is crucial for ensuring the responsible deployment of AI technologies. This will necessitate ongoing collaboration between regulators, industry stakeholders, and researchers.

Finally, the ethical considerations surrounding AI in finance will continue to be a focus of attention. The potential for AI to exacerbate existing inequalities or create new ones demands a proactive and responsible approach. The development of ethical guidelines and standards for the use of AI in finance will be essential in ensuring its responsible and equitable deployment.

In conclusion, this book has provided a comprehensive overview of the application of AI in finance, exploring both its transformative potential and the associated challenges. The future of finance is inextricably linked to the responsible development and deployment of AI technologies. By understanding the fundamental principles, navigating the ethical considerations, and proactively managing the inherent risks, we can harness the power of AI to create a more efficient, transparent, and equitable financial system. The journey has just begun, and the collaborative efforts of technologists, financial professionals, and regulators will be essential in shaping a future where AI empowers and benefits all stakeholders in the financial ecosystem. The responsible adoption of AI in finance will be critical not only for individual success but also for the stability and prosperity of the global financial system as a whole. This requires continuous learning, adaptation, and a commitment to ethical practices. The insights shared throughout this book provide a foundation for navigating this exciting and rapidly evolving landscape.

The integration of artificial intelligence (AI) into the investment world is no longer a futuristic concept; it's a rapidly unfolding reality, reshaping how investments are made, managed, and understood. This transformative power stems from AI's ability to process and analyze vast quantities of data far exceeding human capacity, identifying patterns and insights that would otherwise remain hidden. This allows for more informed, efficient, and potentially lucrative investment strategies.

One of the most significant impacts of AI is on algorithmic trading. High-frequency trading (HFT) algorithms, powered by AI, can execute thousands of trades per second, capitalizing on minuscule market fluctuations that would be impossible for human traders to detect and react to. This speed and scale provide significant advantages, but also raise concerns about market stability and the potential for manipulation. Sophisticated AI algorithms can identify and exploit arbitrage opportunities with unprecedented speed and efficiency, leading to substantial profits for those employing them. However, the opacity of some of these algorithms, their "black box" nature, necessitates the development and implementation of robust regulatory frameworks to ensure fair market practices and prevent the concentration of power in the hands of a few sophisticated players. The development of explainable AI (XAI) is crucial in this regard, allowing for greater transparency and scrutiny of these powerful trading systems. Furthermore, the risk of cascading failures in interconnected HFT systems needs careful consideration and mitigation strategies. While AI-driven HFT promises efficiency and speed, the potential for systemic risks necessitates careful monitoring and regulation.

Beyond algorithmic trading, AI is revolutionizing portfolio management. Traditional portfolio construction often relied on simplified models and historical data, limiting its ability to adapt to rapidly changing market conditions. AI algorithms, on the other hand, can analyze a much wider array of data points, including alternative data sources such as social media sentiment, news articles, and satellite imagery, providing a more holistic and dynamic view of market trends. This allows for personalized investment strategies tailored to individual investor risk profiles, financial goals, and time horizons. AI can not only optimize portfolio construction but also dynamically adjust the portfolio based on real-time market changes, potentially maximizing returns and minimizing risks. This dynamic approach to portfolio management offers a significant advantage over traditional static methods, particularly in volatile markets. For example, AI can identify early warning signs of market downturns by analyzing news sentiment or economic indicators, allowing investors to adjust their portfolios proactively.

Risk management, a cornerstone of successful investing, is another area profoundly impacted by AI. AI-powered systems can analyze vast datasets to identify potential risks, including systemic risks, credit risks, and operational risks, with greater accuracy and speed than traditional methods. This allows for more proactive risk mitigation strategies, minimizing potential losses. AI algorithms can identify subtle correlations and patterns that may be overlooked by human analysts, providing early warnings of impending problems. For instance, AI can analyze credit scores, loan applications, and market trends to predict loan defaults with greater accuracy, enabling lenders to manage their credit risk more effectively. Moreover, AI can help identify and mitigate fraud, a major concern in the financial world. Sophisticated algorithms can detect unusual patterns in transactions, flagging potentially fraudulent activities for further investigation. This enhances security for both investors and financial institutions. The application of AI in risk management is not about replacing human expertise but enhancing it, allowing human analysts to focus on more complex and strategic issues while AI handles the more routine tasks of data analysis and risk identification.

The accessibility of investment opportunities is also significantly enhanced by AI. AI-powered robo-advisors offer automated investment management services, making professional investment advice accessible to a wider range of individuals, including those with limited financial expertise or resources. These robo-advisors can provide personalized investment strategies based on individual circumstances, significantly lowering the barrier to entry for many potential investors. This democratization of investment opportunities has the potential to improve financial inclusion and economic well-being for a wider segment of the population. Furthermore, AI-powered platforms can provide personalized financial education and guidance, helping investors understand the intricacies of investing and make informed decisions. This enhanced accessibility through AI-powered tools promises a more equitable and participatory financial ecosystem.

However, the integration of AI in investing isn't without challenges. Algorithmic bias, stemming from biased training data, remains a significant concern. If the data used to train AI algorithms reflects existing societal biases, the resulting investment strategies may perpetuate and even exacerbate those biases. This can lead to unfair or discriminatory outcomes, particularly for underrepresented groups. Therefore, careful attention must be paid to data curation and algorithm development to mitigate algorithmic bias. Regular audits and monitoring of AI systems are essential to identify and address any signs of bias. Transparency in algorithms is also crucial; explainable AI (XAI) is necessary not only for regulatory compliance but also to ensure that investors understand how decisions are being made and the potential risks involved. Furthermore, the potential for market manipulation through sophisticated AI-driven strategies requires vigilant regulatory oversight.

The future of AI in investing hinges on responsible development, deployment, and regulation. Continuous advancements in machine learning and other AI technologies promise further improvements in efficiency, accuracy, and accessibility. However, these advancements must be accompanied by a strong ethical framework and robust regulatory oversight to ensure fairness, transparency, and market stability. The development of clear guidelines for the use of AI in investment management, addressing issues such as data privacy, algorithmic bias, and market manipulation, is crucial. Collaboration between industry stakeholders, regulators, and researchers is essential in shaping a future where AI empowers both investors and the financial system as a whole. The potential for AI to enhance efficiency, personalize investment strategies, and democratize access to investment opportunities is immense, but responsible governance and ethical considerations are paramount to realizing this potential while mitigating the associated risks. The ongoing dialogue and collaboration between technologists, financial professionals, and regulators are crucial in navigating the complex landscape of AI in finance and ensuring its positive impact on the global economy. The path forward requires not only technological innovation but also a commitment to ethical principles, transparency, and responsible regulation.

The integration of AI into finance, while revolutionary, is far from a solved problem. Numerous challenges remain, demanding ongoing research and development to fully realize the technology's potential while mitigating inherent risks. A crucial area requiring attention is the refinement of AI algorithms themselves. Current algorithms, even the most sophisticated, often rely on historical data, which may not accurately reflect future market behavior. The inherent limitations of relying solely on past performance are amplified when dealing with unpredictable events like geopolitical crises, pandemics, or sudden shifts in consumer behavior. Developing algorithms capable of adapting more dynamically to unforeseen circumstances is crucial. This requires moving beyond purely data-driven models to incorporate elements of expert judgment and scenario planning. Hybrid models that combine the strengths of AI with human expertise could provide a more robust and resilient approach to investment management.

Furthermore, the issue of algorithmic bias continues to pose a significant challenge. Even with meticulous data cleaning and preprocessing, biases can seep into AI models, leading to unfair or discriminatory outcomes. This is particularly concerning in areas like credit scoring and loan applications, where biased algorithms could exacerbate existing inequalities. Addressing this requires not only technical solutions—such as developing more robust bias detection and mitigation techniques—but also a broader societal effort to address the root causes of bias in the data itself. Increased diversity in the teams developing and deploying AI algorithms is vital to ensure a more equitable outcome. Rigorous auditing and independent validation of AI systems are essential to identify and rectify biases before they cause significant harm. Transparency in the algorithms used is also crucial, allowing for greater scrutiny and accountability. The development of explainable AI (XAI) remains a significant area of focus, enabling stakeholders to understand the reasoning behind algorithmic decisions and identify potential biases more readily.

The "black box" nature of many complex AI models is a significant hurdle to overcome. While AI's ability to identify complex patterns is impressive, the lack of transparency can hinder understanding of the decision-making process. This opaqueness can erode trust, particularly among regulators and investors concerned about the potential for manipulation or unforeseen consequences. The push for greater transparency in AI algorithms, particularly those used in high-frequency trading, is critical for maintaining market stability and investor confidence. Regulatory bodies are increasingly emphasizing the need for explainable AI, demanding that algorithms be designed to provide insights into their decision-making processes. This is not simply a matter of regulatory compliance; it's essential for fostering responsible innovation and ensuring that AI is used ethically and effectively within the financial ecosystem.

Another significant challenge is data privacy and security. AI models require massive amounts of data to function effectively, and this data often includes sensitive personal information. Protecting this data from unauthorized access and misuse is paramount. The implementation of robust cybersecurity measures and compliance with data privacy regulations, such as GDPR and CCPA, is essential. Furthermore, the potential for data breaches and cyberattacks on AI systems poses a significant risk, demanding ongoing investment in cybersecurity infrastructure and expertise. Building secure and resilient systems is critical to safeguarding both investor data and the stability of the financial system as a whole. The ethical implications of using personal data for AI-driven investment strategies also require careful consideration, ensuring that data is used responsibly and transparently, respecting individual privacy rights.

The regulatory landscape surrounding AI in finance is still evolving, creating both challenges and opportunities. The lack of clear and consistent regulations can hinder innovation, while overly restrictive regulations could stifle the development of potentially beneficial AI applications. A balanced approach is needed, fostering innovation while ensuring appropriate oversight and protection of investors. International collaboration among regulators is vital to develop consistent standards and prevent regulatory arbitrage. Clear guidelines are needed for data usage, algorithm transparency, and risk management, addressing issues such as algorithmic bias and market manipulation. Regulatory frameworks need to be adaptable and forward-looking, capable of keeping pace with rapid technological advancements.

Despite the challenges, the future outlook for AI in investing is undeniably promising. Advancements in machine learning and deep learning are expected to lead to even more sophisticated and accurate algorithms, capable of analyzing increasingly complex datasets and identifying previously undetectable patterns. This could lead to significant improvements in portfolio optimization, risk management, and fraud detection. The potential for personalized investment advice and access to a wider range of investment opportunities is also considerable. AI-powered robo-advisors are expected to play an increasingly significant role in making investment management more accessible and affordable for individual investors.

The development of more sophisticated AI models capable of incorporating alternative data sources, such as social media sentiment, news articles, and satellite imagery, offers exciting possibilities for market analysis and prediction. This integration of diverse data sources can provide a more holistic and nuanced understanding of market trends, potentially improving investment strategies and risk management. The use of AI in identifying and mitigating systemic risks is also a crucial area of development. Sophisticated AI algorithms could play a vital role in identifying early warning signs of financial crises and helping policymakers develop effective responses.

The development of hybrid models, combining AI's computational power with human expertise and judgment, holds significant potential. Such models could leverage the strengths of both approaches, mitigating the limitations of each.

Human analysts can provide valuable contextual knowledge and intuition, while AI algorithms can handle the vast amounts of data analysis required for effective investment decision-making. This collaborative approach can lead to more robust and resilient investment strategies.

Furthermore, the ongoing development of XAI is critical for ensuring trust and transparency. As AI algorithms become increasingly complex, the need for explainability becomes even more acute. XAI techniques aim to make the decision-making processes of AI models more transparent and understandable, allowing investors and regulators to assess the risks and benefits more effectively. This increased transparency will be crucial for building trust and ensuring the responsible use of AI in finance.

The future of AI in investing will hinge on a collaborative effort between technologists, financial professionals, regulators, and researchers. The development of ethical guidelines, robust regulatory frameworks, and effective educational initiatives is vital to ensure that AI is used responsibly and ethically. A focus on transparency, accountability, and fairness is essential to ensure that the benefits of AI are shared broadly and that the risks are effectively mitigated. Ongoing research and development are needed to address the challenges and unlock the full potential of AI in finance, creating a more efficient, transparent, and equitable financial ecosystem. This journey requires a continuous dialogue and a shared commitment to responsible innovation, ensuring that AI serves as a force for good in the global financial landscape. The ethical considerations and responsible governance surrounding AI's integration into finance will ultimately determine its long-term success and its ability to improve the lives of individuals and the stability of the financial markets worldwide.

The preceding chapters have explored the transformative potential of artificial intelligence (AI) in finance, detailing its capabilities and the challenges it presents. We've examined the intricacies of AI algorithms, the crucial need for addressing algorithmic bias, and the importance of transparency and explainability. We've also delved into the complexities of data privacy, cybersecurity, and the ever-evolving regulatory landscape. Now, it's time to move beyond analysis and embrace the call to action: actively participate in the AI revolution reshaping the financial world.

This isn't a call to blindly adopt every AI-powered tool that comes along. Instead, it's an invitation to engage critically, thoughtfully, and strategically. The key is to understand the strengths and limitations of AI and to integrate it into your investment strategies in a way that enhances, not replaces, human judgment and intuition. This requires a paradigm shift, a move away from purely reactive strategies to a proactive, data-driven approach informed by the insights AI can provide.

The first step towards embracing this revolution involves continuous learning. The field of AI is rapidly evolving, with new algorithms, techniques, and applications emerging constantly. Staying abreast of these developments is not merely beneficial; it's essential for remaining competitive and mitigating potential risks. This means actively seeking out high-quality educational resources, participating in industry conferences, and engaging with experts in the field. Online courses, webinars, and professional development programs offer numerous avenues for enhancing your AI literacy. Building a strong network of contacts within the fintech community can also prove invaluable, facilitating access to the latest insights and best practices.

Beyond formal education, a proactive approach to learning involves experimenting with different AI tools and technologies. Many platforms now offer user-friendly AI-powered tools designed for both individual investors and institutional players. Experimenting with these tools – from robo-advisors to sentiment analysis platforms – allows for a practical understanding of their capabilities and limitations. This hands-on experience is invaluable in developing a nuanced understanding of how AI can be effectively integrated into your investment strategy.

However, experimentation should be approached with caution. It's crucial to start small, focusing on specific areas where AI can add value. Instead of trying to overhaul your entire investment process at once, begin by using AI tools to augment your existing strategies. For example, you might start by using AI-powered tools for market research, identifying potential investment opportunities, or automating repetitive tasks. As your comfort level and understanding increase, you can gradually expand your use of AI, gradually incorporating more advanced techniques..

A crucial aspect of responsible AI integration is risk management. While AI can significantly enhance investment decision-making, it's not a panacea. AI models are only as good as the data they're trained on, and biases can creep into even the most sophisticated algorithms. Therefore, it's essential to thoroughly evaluate the reliability and accuracy of AI-powered tools before incorporating them into your strategy. Consider diversifying your AI sources to mitigate the risks associated with relying on a single algorithm or platform.

Moreover, it's vital to remain aware of the ethical considerations surrounding AI in finance. Algorithmic bias, data privacy, and transparency are crucial issues that require careful consideration. Always ensure that the AI tools you use adhere to the highest ethical standards, respecting individual privacy rights and minimizing the potential for discriminatory outcomes. Prioritize platforms that are transparent about their algorithms and data sources, enabling you to understand the reasoning behind their recommendations.

Beyond individual adoption, the financial industry as a whole needs to embrace a culture of responsible AI innovation. This requires collaboration between technologists, financial professionals, regulators, and researchers. The development of ethical guidelines, robust regulatory frameworks, and effective educational initiatives is paramount. A focus on transparency, accountability, and fairness is essential to ensure that the benefits of AI are shared broadly and that the risks are effectively mitigated. This collaborative approach will be instrumental in shaping the future of AI in finance, ensuring it serves as a force for good in the global financial landscape.

Furthermore, continuous adaptation is crucial. The financial markets are dynamic and complex, and AI models need to be regularly updated and refined to keep pace with evolving market conditions. This necessitates ongoing monitoring and evaluation of the performance of your AI tools, making adjustments as needed to optimize their effectiveness. Regularly review the data sources used by your AI models, ensuring they remain relevant and reliable. As new data sources and AI techniques emerge, be open to experimenting and incorporating these advancements into your strategies.

Embracing the AI revolution in investing is not simply about adopting new technology; it's about transforming your approach to investing itself. It requires a willingness to learn, adapt, and embrace change. It demands a critical and thoughtful approach, balancing the power of AI with the irreplaceable value of human judgment and intuition. By proactively engaging with AI tools and techniques, while remaining mindful of the associated risks and ethical considerations, you can harness the transformative power of AI to enhance your investment strategies and navigate the complexities of the modern financial landscape. The future of finance is interwoven with AI; those who embrace this reality will be best positioned for success.

The path forward isn't without its challenges. Regulatory uncertainty, the potential for algorithmic bias, and the need for enhanced cybersecurity remain significant hurdles. However, by actively engaging in the dialogue surrounding these issues, participating in industry initiatives focused on responsible AI development, and advocating for transparent and ethical practices, we can collectively shape a future where AI serves as a powerful force for positive change in the financial sector. The potential benefits—improved efficiency, increased accessibility, and enhanced risk management—are simply too significant to ignore. The call to action is clear: embrace the AI revolution, but do so responsibly, ethically, and strategically. The rewards for those who navigate this transformation successfully will be substantial, reshaping both personal investment outcomes and the broader financial ecosystem. This journey requires a commitment to lifelong learning, a willingness to adapt, and a deep understanding of the ethical responsibilities inherent in harnessing the power of AI. The future is not just about AI; it's about human ingenuity guided by intelligent technology.

The journey into AI-powered investing is a continuous learning process. To fully grasp its potential and navigate its complexities, ongoing education is crucial. Fortunately, a wealth of resources exists to support this pursuit, offering insights into the latest advancements and best practices. This section serves as a curated guide to these resources, categorized for ease of access and tailored to different levels of expertise.

For beginners seeking a foundational understanding, several online platforms provide excellent introductory courses. Coursera and edX, for example, offer a range of courses on AI, machine learning, and their applications in finance, many of which are taught by leading academics and industry professionals. These courses often incorporate practical exercises and case studies, enhancing the learning experience and translating theoretical knowledge into practical skills. Look for courses focusing on topics such as "Introduction to Machine Learning," "Financial Modeling with Python," and "Algorithmic Trading Strategies," ensuring they are up-to-date and aligned with the current state of the art. Many of these courses are free to audit, allowing you to access the course materials and lectures without paying for a certificate of completion. However, obtaining a certificate can be beneficial for professional development.

Beyond introductory courses, platforms like Udacity and DataCamp provide more specialized training in areas such as natural language processing (NLP) for sentiment analysis in financial markets, and deep learning techniques for predictive modeling. These platforms often offer project-based learning, allowing you to build your portfolio and demonstrate practical skills to potential employers or collaborators. Furthermore, they frequently update their course offerings to reflect the latest breakthroughs in the field, ensuring you are always learning the most relevant and current information.

For those seeking deeper dives into specific aspects of AI in finance, various professional development programs are available. Many universities and specialized institutions offer postgraduate certificates or master's degrees focused on fintech and AI. These programs are ideal for individuals seeking to transition their career into this field or significantly enhance their existing expertise. Look for programs that blend theoretical coursework with practical projects and industry collaborations. Such programs often involve direct interaction with leading researchers and practitioners, providing invaluable networking opportunities and insights into real-world applications. The emphasis should be on programs that focus on the intersection of finance and AI, offering a strong theoretical foundation coupled with practical skill-building.

In addition to formal education, numerous books provide comprehensive overviews of AI and its implications for finance. "AI Superpowers" by Kai-Fu Lee offers a broad perspective on the global implications of AI, while more specialized texts such as "Machine Learning for Algorithmic Trading" by Stefan Jansen provide in-depth technical details relevant to the application of AI in investing. When selecting books, prioritize those written by recognized experts in the field, ensuring they provide accurate and up-to-date information. Also, look for books with a strong emphasis on practical applications, offering concrete examples and case studies. Reading reviews from other professionals can also provide a valuable insight into the book's quality and relevance.

Staying updated on the latest developments requires engagement with industry publications and research papers. Journals such as the

Journal of Financial Econometrics and the *Journal of Banking & Finance* regularly publish research on the application of AI in finance. These academic publications often present cutting-edge findings and methodologies, offering invaluable insights into the newest trends. While these papers may require a strong background in finance and mathematics, they are essential for anyone seeking to remain at the forefront of the field. Additionally, online platforms such as arXiv provide access to pre-print research papers, allowing you to follow the latest discoveries before they appear in formal publications.

Reputable online news sources dedicated to fintech and AI offer valuable commentary and analysis on current events and emerging technologies. Websites such as Financial News, The Banker, and Bloomberg offer insightful articles and expert opinions on the latest developments. These sources can provide timely updates on regulatory changes, technological advancements, and market trends. It's critical to evaluate the credibility of any online source, paying attention to the authors' expertise and any potential conflicts of interest. Remember, not all sources are equal, so rely on established, reputable media outlets.

Furthermore, actively participating in industry conferences and workshops is beneficial for networking and staying abreast of current trends. Conferences such as those organized by the Association for Computing Machinery (ACM) and the Institute of Electrical and Electronics Engineers (IEEE) often include sessions dedicated to AI and finance. Attending these events provides opportunities to learn from leading experts, network with peers, and gain insights into future developments. Networking opportunities are exceptionally valuable, allowing you to exchange ideas, collaborate on projects, and build relationships within the fintech community. Presenting your own research at these conferences can also enhance your reputation and visibility within the field.

Beyond formal publications and conferences, actively engaging with online communities and forums dedicated to AI and finance can enhance your learning and provide opportunities for collaboration. Platforms such as Reddit and Stack Overflow often host discussions on specific aspects of AI implementation, providing a space to ask questions, seek advice, and share insights. Participation in these communities can also expose you to alternative perspectives and unconventional approaches to solving problems. Be aware, however, that the information found in online forums should be evaluated critically, as it is not always vetted for accuracy or completeness. However, they can be a valuable source of diverse viewpoints.

The resources outlined above provide a starting point for your ongoing education in AI-powered investing. It's crucial to develop a personalized learning plan, tailoring your focus to your specific goals and interests. Continuously refining your knowledge, keeping abreast of emerging trends, and engaging with the wider community are essential aspects of navigating the dynamic world of AI in finance. The field is constantly evolving; continuous learning is not simply advantageous—it's essential for sustained success. By actively engaging with the resources available and committing to lifelong learning, you can ensure your continued growth and adaptation within this rapidly evolving landscape. This proactive approach will not only enhance your understanding of AI's capabilities but also equip you to harness its potential for responsible and profitable investment strategies. Remember, the future of finance is inextricably linked with AI. By actively participating in its development and responsible application, you can secure your position at the forefront of this transformative revolution.

AI (Artificial Intelligence): The simulation of human intelligence processes by machines, especially computer systems.

Algorithmic Trading: The use of computer programs to follow a defined set of instructions (an algorithm) to place a trade.

Deep Learning: A subfield of machine learning that uses artificial neural networks with multiple layers to analyze data.

Fintech: A portmanteau of "financial technology," encompassing new tech applications in the financial industry.

Machine Learning: A type of AI that allows software applications to become more accurate in predicting outcomes without being explicitly programmed.

Natural Language Processing (NLP): A branch of AI focused on enabling computers to understand, interpret, and generate human language.

Predictive Modeling: The process of using data and algorithms to predict future outcomes.

Sentiment Analysis: The use of NLP to determine the emotional tone behind a body of text.

Did you love *The AI Investor*? Then you should read *Swift to the Stars*[1] by DaJohn Garrett and Alander Lee Pulliam Jr!

[2]

"Swift to the Stars" is the inspiring story of a young Black rapper named Swift, growing up in the gritty streets of Sylmar, California. With dreams of fame, fortune, and freedom, he constantly battles the cycle of poverty, reckless decisions, and societal expectations. His life is a whirlwind of late-night raps, wild nights with women, and fleeting moments of clarity about a future beyond the ghetto.

Driven by an unshakable desire to escape, Swift begins to plan something extraordinary—traveling to Mars. Despite skepticism from friends, family, and the industry, he secretly studies rocket science, builds his own spacecraft, and prepares for a journey no one believed possible.

From sketching spaceship plans on his couch to launching himself into the cosmos, Swift's story is a testament to perseverance, ambition, and the power of

---

1. https://books2read.com/u/mZMdpl

2. https://books2read.com/u/mZMdpl

dreaming big. His journey underscores that no matter where you come from, your future is written by the courage to chase the impossible.

Also by Alander Lee Pulliam Jr

Breaking Free From Foolish Living
Minding Your Own Business

Cyber War and Secret Agencies
Shadow Protocols

Hustlers Ambitions
Undercover Hustle

Society Falls
Unlawfully Detained: Journey to Justice
Echoes of Tomorrow
Rise of The Independents

The Ultimate Creator's Legal Playbook
NextGen Rights

Standalone

The Chronicles of Yen: Unrevealed Prophesy
Breaking Away From The Status Quo
NAVIGATING THE ENTERTAINMENT INDUSTRY: Artificial Intelligence
Ghetto Grace: Breaking Through Strongholds
My Spot Ain't Yours: Stop Hating
Woven Parables of Wisdom
Can't Sway Me
Insecure Hearts
Investing In The New Age of Tech
Mindset Without Agenda: Embracing Authenticity
The Double-Edge Algorithims
The Heart That Gave
Fool's Paradise Wisdom's Reality
The Future of Acting In the Digtial Age
Future In Flames
My Curves Straight
Swift to the Stars
The AI Investor

Also by Olom "Nosa" Okonta

The AI Investor

About the Author

Alander Pulliam is an american film producer,author, actor, writer, business owner and investor. Running for President of the United States 2024. Forming many corporations throughout his career, spending many years in the major label spectrum, he was responsible for several successes in the music industry. In 2015, Alander Pulliam was chosen as one of the Top 25 Marketing Innovators in the World. Serving as an Executive in the Entertainment Industry. Working on TV Shows and Movies, from Producing to Investing. He has made his mark in the industry. Now developing the ABM Streams APP which allows anyone to make money streaming anything and everything that can be streamed. Alander Pulliam is continuing to develop several empires throughout the world of Entertainment. Helping others establish themselves. Working with many of today's artist along with several of the most successful artist coming on the scene in the mid 2000's era of music. Serving as a major label executive, he gradually moved into the TV and Film spectrum.

About the Publisher

Alander Pulliam is an accomplished American film producer, author, actor, writer, entrepreneur, and investor. He is currently a candidate for President of the United States in the 2024 election. Throughout his career, he has founded numerous corporations and has extensive experience within the major label music industry, where he contributed to several notable successes.

www.ingramcontent.com/pod-product-compliance
Lightning Source LLC
LaVergne TN
LVHW090321210926
848480LV00001B/324
* 9 7 9 8 2 3 0 3 8 8 2 9 6 *